MW00378156

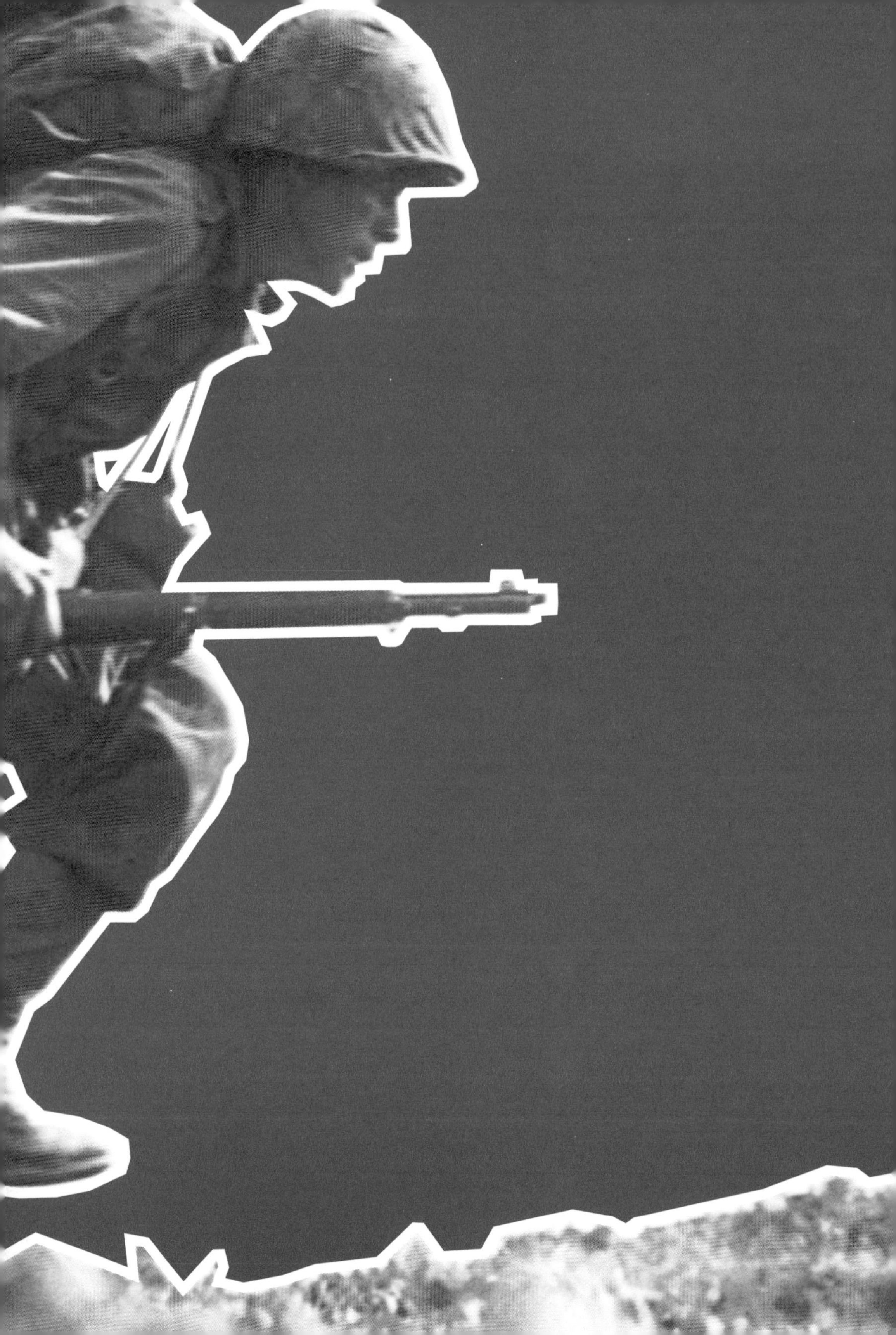

ROB JOHNSON • MICHAEL WHITBY • JOHN FRANCE

HOW TO WIN ON THE BATTLEFIELD

25 KEY TACTICS TO OUTWIT, OUTFLANK AND OUTFIGHT THE ENEMY

with 113 illustrations, including 28 battle plans

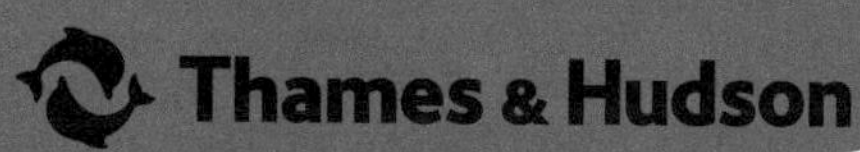

This book is dedicated to Lt S. K. Horn, USMC, and Lt H. Levy, USAF, and all their comrades in the United States Armed Forces

Half-title and title page: A US Marine dashes through Japanese machine gun fire at Okinawa, 10 May 1945.

First published in 2010 in hardcover in the United States of America by Thames & Hudson Inc., 500 Fifth Avenue, New York, New York 10110

thamesandhudsonusa.com

Library of Congress Catalog Card Number 2009936404

ISBN 978-0-500-25161-4

Printed and bound in China by 1010 Printing International Ltd

Contents

Contents

Preface

Throughout history, military leaders have faced the dilemma of formulating the right tactical plan to achieve victory on the battlefield. Much depends on the preparations for that moment: having the correct force structures and strength, appropriate levels of training and morale, or assembly into the most advantageous positions. However, when the fighting begins, commanders must seek to apply the forces at their disposal in a decisive way.

In this book we examine and then analyse, through a series of case studies, 25 of the most important tactics used in war on land, at sea and in the air. These are illustrated with examples from around the globe and from across history in order to demonstrate the enduring value of each one. While much of the current intellectual debate on war focuses on what is new, or even 'post-modern', we are eager to show that, despite obvious changes in weaponry, technology and mobility, certain key techniques and principles have been applied with equal success in the military encounters of the past and are still relevant today. Since so many works of military history focus narrowly on the West, we have selected examples from as wide a field as possible and we have tried to raise awareness of some less well-known battles. Our intention is to highlight the function of each tactical concept in its historical setting.

As authors, we have combined each of our areas of expertise and academic research, and so Rob Johnson has written the sections on early modern and modern warfare, Michael Whitby has offered contrasts and comparisons with examples from the ancient period and John France has focused on battles from the medieval era.

Rob Johnson, Michael Whitby and John France

Winning on the Battlefield: The Aims and Principles of War

What are the most successful tactics of war that have been employed on the battlefield through the ages? And what can we learn from them today? Tactics are the short-term plans and the techniques used to overcome the enemy, usually when in close contact, including the skills and manoeuvres that best deploy one's own forces. Tactics fit within the framework of a strategy, which is a long-term plan or guiding idea, involving the management and manoeuvre of forces and assets in a campaign or war, focused to achieve success. This book explores the concept of tactics through a series of case studies.

Military history cannot provide formulas for guaranteed success in modern war, but an examination of the techniques and manoeuvres of commanders in the past reveals patterns and principles that can be instructive. A close study of the great theorists of war can also be illuminating. Perhaps the earliest known is the ancient Chinese strategist Sun Tzu, who is credited with being the author of the influential and still widely read *The Art of War*, though his dates – and even his existence – are uncertain. Two of the most famous theoreticians of war were both connected, in different ways, with Napoleon. Antoine-Henry, later Baron de, Jomini (1779–1869) fought in the Napoleonic armies and was present at the Battle of Austerlitz. The Prussian Carl von Clausewitz (1780–1831), whose most famous work is translated as *On War*, served throughout the Revolutionary and Napoleonic wars and was profoundly influenced by the experience of fighting against Napoleon.

A late 14th-century manuscript illumination of the Battle of Bouvines: the bold flanking attacks and disciplined action of the French cavalry produced victory for Philip Augustus.

The Principles of War

Baron de Jomini suggested in fact that there could be no theory of strategy, and that to achieve battlefield victory commanders' decisions had to be based on a combination of intuition, experience and an understanding of the principles of war.[1] The most common and important precepts, both on and off the battlefield, can be summarized as:

The Objective A clear, coherent policy or aim is the first priority. This also necessitates the correlation of the ends and the means so that a military victory coincides with the achievement of a political objective.

Offence Victory is dependent on seizing and retaining the initiative by offensive action, but it also requires the clear identification of the enemy's centre of gravity and his defeat there (see **1**). Even when forced into a defensive stance, every opportunity must be sought to off-balance the enemy and regain the initiative, since it is very difficult to achieve victory in a purely reactive posture.

Co-operation The co-ordination of allies and objectives, arms and logistics is a key component for success.

Concentration of force Clausewitz, perhaps the best-known Western strategist, argued that concentration of force was the single most important principle of war, as it ensures local superiority at a critical geographical point (see **13**). At the higher level it also implies that fighting more than one enemy at a time should be avoided, while also seeking to divide the enemy's forces.

Previous pages: Alexander the Great understood that all manoeuvres had to be carried out with bold resolution; this, and his personal drive, secured for him one of the world's greatest empires.

Economy of force The effective distribution of limited resources requires good management, but this is especially important if a commander is to achieve victory in a short, decisive conflict rather than risk a long, drawn-out war of attrition. The exception to this is the strategy of the insurgent, which demands that personnel are conserved while the enemy is exhausted over an extended period.

Manoeuvre, surprise and deception To Sun Tzu the highest art of warfare was to produce a victory without recourse to arms at all, and this could be achieved by a combination of manoeuvre, deception and good intelligence. Manoeuvring into the right position makes possible all the other principles.

Security Good security is essential in order to avoid being deceived, but it is also imperative to be able to ascertain the strengths and weaknesses of the enemy, his movements and even his likely intentions. Conflict between polities is inevitable because of opposing interests, but at the higher level, some wars or battles can be avoided by security, diplomacy, deception, deterrence and manoeuvre.

Simplicity Given the complexity of conflicts, and the need to react to, as well as initiate, operations, simple plans have a better chance of success.

In his writings Carl von Clausewitz captured the essence of war as it existed in his own time, but his ideas still resonate today.

Clausewitz: On War

A pragmatic but thoughtful guide to battlefield operations can be found in the writings of Carl von Clausewitz.[2] His work, which was published after his death, is complex and can sometimes appear to be contradictory. However, it is clear that Clausewitz was eager to analyse war at a far deeper level than the simple tactics and manoeuvres of his own era. He argued that war is essentially violence without natural limits, and it follows from this that the only purpose of war is to bring about combat. War, he famously observed, is an extension of the political will with an admixture of other means, and asserting the will of an army is the purpose of a campaign. Delay, deception and ruses can often be a waste of time, since they do not bring about a decisive conclusion.

Clausewitz was aware that, despite these simple basic elements, war is still the realm of chance. Since war is a collision of wills, it is rarely decided by a single blow. It always occurs in an environment of 'friction' or a 'fog', so that carefully laid plans can be upset by changes in the enemy's decisions and actions, the weather or even just by accidents. Yet there were more predictable patterns too. No doubt mindful of Napoleon's disastrous invasion and retreat from Russia in 1812 (see p. 182), Clausewitz posited that the attack weakens as it penetrates deeper into enemy territory, and, unless arrested in some way, it reaches a culminating point where it is reduced to the same strength as that of the defender. However, identifying this moment is extremely difficult, requiring the judgment of an outstanding general. Clausewitz was certainly conscious that exceptional leaders could have a decisive influence on war. He regarded Napoleon as a military genius, and felt that while it was not

Napoleon on campaign in 1814: his victories were the result not just of his judgment and speed of manoeuvre, but also his ability to understand what motivated men in war and peace.

possible for such attributes as he possessed to be 'trained' in others, they could be encouraged and developed. Conversely, he recognized that all men have their limits – their morale can affect a campaign or a battle, and they are often subject to fatigue, fear and failure. He also acknowledged the particular difficulty faced by regular, professional troops when confronted by an insurgency.

Clausewitz regarded war as a 'remarkable trinity', an interplay of violence, chance and the rational, intellectual and political purpose. The appeal of Clausewitz is his ability to articulate this combination so effectively. He provided no checklist for victory, but he did clarify the purpose of warfare: to 'disarm' the enemy and render him politically helpless and militarily impotent through combat. He made the point that leaders should consider what type of conflict they are engaged in, and therefore what resources should be committed to it. He also thought that generals should be allowed freedom of action once war has broken out, without political interference. He advocated the building of a wide coalition and the training and preparation of troops before fighting begins. Above all, he stressed the need for the determination to win to be inculcated into all. Such a determination ensures that one's forces are willing to take casualties and still strive to achieve the objectives set out for them. He stated that if one's own side has more resolve and more resources, it should advance and employ the maximum force available at the outset in order to achieve victory in the shortest time. This result is most likely to come about, he argued, by destroying the enemy's army in battle – the centre of gravity.

The Clausewitzian approach to achieving battlefield victory is followed closely in this book. His key principles of the offensive, concentration and continuity, the value of recognizing the friction of war yet still planning with flexibility, as well as his other ideas, underpin what is set out here. However, while acknowledging his criticisms of some of the peripheral aspects of war – such as his doubts about the usefulness of manoeuvre, deception and intelligence – these tactics are incorporated because they have, on occasion, also proved decisive.

The Utility of Experience

Clausewitz noted that no wars are alike, and there are inherent risks in trying to present comparisons across wide historical periods. Yet there is some merit in being able to link theoretical ideas in this way, in order to test their validity and demonstrate their universality. The aim here is not to compare the battles, but the concept.

Many more techniques could have been included, but the 25 tactics examined here have been selected for their importance and variety. The first five deal with the attack and counter-attack, or manoeuvres specifically associated with them. The historical examples show how some of these operational concepts were arrived at after a great deal of reflection, as in the case of Montgomery's Battle of El Alamein (1942), while others, such as Robert E. Lee's bold flanking march at Chancellorsville (1863), came about through hasty but brilliant inspiration. Tactic **6** acknowledges how the environment can play a role in victory if used to advantage, while the concepts in **7**, **8** and **9** all illustrate how assets can be deployed to maximum effect. The focus in **10** to **13** is manoeuvre, whereas **14** and **15** are concerned with the psychology of seizing the initiative and off-balancing the enemy. Tactic **16** highlights the value of weight and numbers, while **17**, **18**, and **19** provide indications of how a tactical defence can be conducted to produce victories, or at least the opportunities for them. In **20** and **21** we return to the issue of psychology, dealing specifically with deception and terror. Attrition as a tool to victory is analysed in **22**, and **23** sums up the contribution that can be made by intelligence. The final two concepts bring the book up to date, with insurgency and counter-insurgency.

There is, of course, in a work such as this, the danger of a general overemphasis on manoeuvre on the battlefield, when wars are often decided by other means. Sieges and naval campaigns, for example, have frequently produced strategic victories without the drama of battles. Alliances and coalitions have always been important to the outcome of a conflict, but individual national histories have a tendency to downplay or even ignore this critical dimension. Moreover, all the operations referred to in this book, while having some common principles and points of comparison, were conducted under vastly different conditions and it would be misleading to give the impression that battles have always been the sole arbiters of victory.

The *Scharnhorst*, the technological epitome of the battleship in its day, was nevertheless vulnerable to carrier-borne aircraft and torpedoes.

The tactical concepts that have been chosen for analysis here are ones that have remained important, despite evident changes in technology through history. It is always easy to overestimate the impact of technology as a determinant of victory. After the First World War, it was thought that air power alone would allow states to reduce, and therefore dominate, the cities of their enemies, and perhaps break their morale through bombing.[3] Yet the use of air power on its own has not proved a successful strategy – ground and naval forces are needed in combination. Major General J. F. C. Fuller, a First World War tank officer and military historian, developed the first armoured doctrine in the inter-war years, which was embraced enthusiastically by the German army, but he believed that armour still had to act in concert with the other arms. Indeed, the Second World War indicated that all new technologies had to be built into the context of combined operations to be successful. No single *Wunderwaffe* ('wonder weapon') could provide a guarantee of victory.

After the 1940s, it was assumed that nuclear weapons, given their devastating effects, had 'absolute' strategic value. However, those powers that possess nuclear weapons have not been immune from war, although they have probably been protected from a total war. A study of the impact of technology on war can produce convincing moments of change, but closer inspection of historical examples suggests that a reliance on a new technology on its own was not sufficient to produce a victory, and this is as true of the most

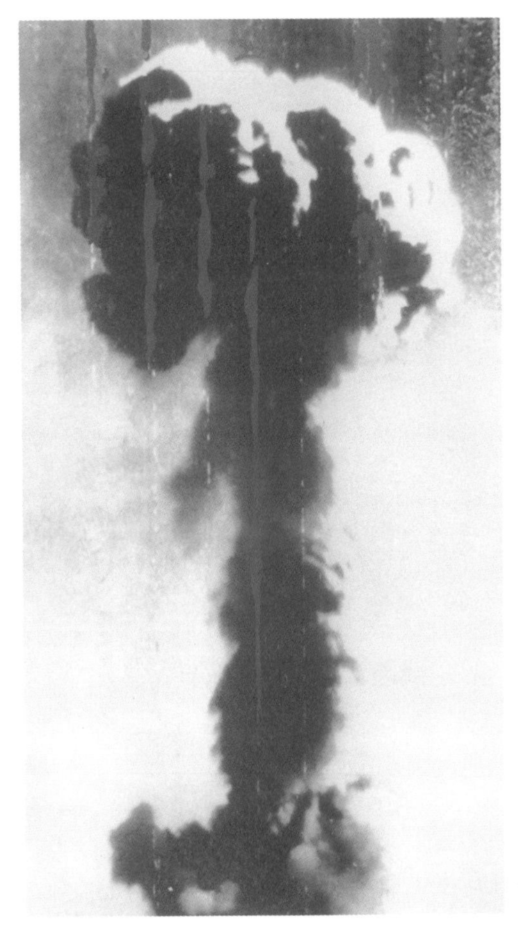

Mushroom cloud from a Chinese nuclear test, 1964. Nuclear weapons have not prevented more conventional wars, but perhaps have averted the conflagration of total war between superpowers.

devastating weapon systems of all. That said, a wholly different scenario would be produced by non-state actors, such as terrorists, using nuclear weapons, or other radiological, chemical or biological weapons of mass destruction.[4] The intelligent strategic use of such a weapon would render a nuclear counter-attack impossible. Putting aside the moral ramifications of such an attack, from a military point of view it simply reflects the need for any belligerent to find and exploit the weakness of his enemy.

This highlights more generally the problem facing tacticians dealing with asymmetrical conflicts – that is those conflicts where strong conventional forces are worn down by the harassing tactics of a weaker opponent. The key issue here is the difficulty of bringing the power of conventional forces to bear on an enemy operating among a civilian population, the destruction of which would be politically unacceptable internationally. Mao Zedong in China and General Vo Nguyen Giap in Vietnam understood the need to flow among the people and to win them to their cause in an insurgency or civil war.[5] By contrast, the United States in Vietnam and the Soviet Union in Afghanistan were frustrated in their attempts to build a political system in their own image, despite military superiority. The Americans in Vietnam won almost every battlefield encounter, but that did not bring them victory in the war. They had not been able to correlate ends and means, the rational calculus of war. Fuller advised his American counterparts that their bombing campaign in Vietnam would not defeat communism, because 'World War Two should have told them that ideas cannot be dislodged by TNT'.[6]

Despite a 'Revolution in Military Affairs' (the dramatic acceleration in the evolution of communications, surveillance, firepower and mobility in the late 20th century), which has given the United States a technological edge and a dominance in almost every spectrum of warfare from land operations to the information/cyber warfare environment, and while distance and topography are no longer barriers to its forces, there have been severe difficulties in securing victory in the aftermath of the invasion of Afghanistan in 2001 and Iraq in 2003. These examples remind us that battlefield superiority may be illusory and that tactical principles, identified throughout history, will still apply even if the force structures that implement them change radically through time. Historically, the ability to maintain flexibility, to learn from mistakes and to develop trends or ideas on how to approach war in the future (rather than trying to 'fight the last war') have been critically important. We hope that, if only on that basis, this book will provide some interesting ideas worthy of further thought.

1

El Alamein, 1942

The Attack at the Centre of Gravity

Clausewitz, the great strategic thinker, was always clear that the route to victory is through the offensive. Invariably, a force that finds itself on the defensive will eventually have to take the offensive and carry the fighting to the enemy. Even in a guerrilla warfare setting (see **24**), strategists such as Mao Zedong and General Vo Nguyen Giap recognized that the final phase of any armed struggle would necessitate forming a conventional offensive force to take control of the country. Remaining on the defensive may, of course, be a means to inflict considerable losses on the enemy and allow a militarily weaker army to sustain itself against a stronger foe, but the offensive gives an army the initiative and allows a commander the chance to impose his will on the course of the campaign.

Field Marshal Bernard Montgomery advocated taking an offensive stance, even when forced on to the defensive.[1] He also understood that an offensive against a strong position required careful planning, in particular to locate and then defeat the enemy at what Clausewitz called the 'centre of gravity' – the precise point and moment where a battle would be decided.[2] For Clausewitz, the centre of gravity was usually in the strength of the enemy's army, although it could also be a geographical location on the battlefield.

Previous pages: A British tank advancing during the Battle of El Alamein. Armour combined with tenacious infantry, artillery and air assaults, gave the Allied offensive momentum.

A good example of this concept can be found in the North African Campaign during the Second World War (1939–45). After British, Imperial and Commonwealth forces had ejected the Italian forces from much of Mussolini's new 'Roman Empire' in East Africa and driven them back through Libya in 1940,[3] the Germans augmented Mussolini's wavering troops with an 'Afrika Korps'. Twice (in spring 1941 and summer 1942) they pushed the British and their allies back towards Egypt. Each side discovered that it could exploit the expanse of desert to make wide flanking manoeuvres to the south, but also that the chief difficulty lay in supply. Although armoured and mechanized formations achieved successes, the further they drove from sources of fuel and ammunition, the harder it became to sustain an offensive. Clausewitz referred to a 'culminating point' in attack (see **13**), a point beyond which the assailants are weakened or overstretched: the Desert War clearly demonstrates this factor.

El Alamein, 1942

In mid-1942, the German and Italian forces under Field Marshal Erwin Rommel had pressed the British and their allies back on to the Egyptian border. The Allies were concerned that the Germans might make a bid to advance into the Middle East, perhaps in conjunction with Axis forces in Russia, to take possession of vital oil supplies. It was imperative that Rommel was halted. The Germans were at the limit of their supply line, which stretched back hundreds of miles to Benghazi and Tobruk, and in July 1942 this helped General Claude Auchinleck to hold the Egyptian border in the First Battle of Alamein. Montgomery was then appointed by the British Prime Minister, Winston Churchill, to attack and defeat the Germans decisively. This would allow Britain to open a new front against Germany in the Mediterranean, thereby dividing Hitler's forces. Montgomery first checked Rommel at the Battle of Alam Halfa (August 1942); it was now time to go on to the offensive.[4]

Field Marshal Bernard Montgomery, who made a relentless series of attacks to break the Axis forces in North Africa at El Alamein.

Montgomery planned his attack carefully, but knew that he would have to remain flexible once the action had begun. He concentrated first on the training, refitting and reinforcement of his army. New, more reliable tanks – the American Sherman – were delivered, and a vast number of guns deployed. Allied air forces gained air superiority, while the Royal Navy intercepted and destroyed an increasing volume of Rommel's supply vessels. Montgomery knew that in this case the centre of gravity was the German 'Panzerarmee' force itself, and he would have to make use of his superior numbers to break the Afrika Korps if he was to win.[5]

Rommel in turn realized that he would have to face this onslaught and had prepared his positions carefully. Critically short of petrol, he knew that he would only be able to make one decisive counter-move with his armoured formations. It was vital that he identified where the main attack would fall and he therefore had to slow the momentum of the attackers. To achieve this he depended on fixed positions: German and Italian infantry created minefields of great depth – as much as 5 miles (8 km) – covered by artillery and hedged with hundreds of miles of barbed wire. These 'Devil's Gardens' would allow the Axis forces to inflict catastrophic casualties, but permit the counter-attacks Rommel envisaged at the *Schwerpunkt* ('critical point'). To the south, his flank was fixed by the great Qattara Depression, which was impassable to Allied armour. The battle would be decided on a front just 37 miles (60 km) wide.

Field Marshal Erwin Rommel, the 'Desert Fox', was defeated at El Alamein, despite a carefully planned defence.

Montgomery's plan was to make a strong feint against the southern flank with elaborate deception measures (including dummy vehicles, depots and a mock fuel pipeline), while making the main attack, known as Operation Lightfoot, against the northern portion of the Axis lines. The assault was to be preceded by the largest artillery bombardment by British forces since the First World War. This would cover the advance of four infantry divisions of XXX Corps as they picked their way through the minefields and broke into the German positions. Once secured, the next phase would involve an armoured thrust by X Corps into the heart of the German lines (the 'dogfight' as Montgomery styled it) and then a break out, striking deep behind their army.[6]

The Break-in: 'Bite and Hold'

The Allied attack got underway as planned at 2140 hrs on 23 October 1942 with a hurricane of artillery fire against the Axis positions, all of which had been identified in advance. Pinned in their entrenchments, the Germans were initially unable to prevent the British, Australian, New Zealand and South African infantry inching their way through the minefields. However, there were a number of problems for the Allies. New mine detectors were found to be too sensitive to be useful, and the units equipped with them had to resort to the traditional method of probing the sand with a bayonet. As they neared the German positions they came under a greater volume of fire, and losses mounted steadily. Smoke and dust raised by the bombardment added to the confusion so that, despite the arc of tracers and pre-positioned lights, it became increasingly difficult to maintain direction towards the objectives. The 51st (Highland) Division troops found themselves ahead of their creeping curtain of artillery fire and suffered casualties as a result. On the left flank, the 1st South African Division was also badly cut up by German fire and had to repel a spirited counter-attack before doggedly fighting into the enemy trenches. In the centre, there was a bitter close-quarter battle for the Miteirya Ridge, but the Australians and New Zealanders took all their objectives.

Behind the infantry, who had been slowed almost to a standstill, the British armour was struggling to get through the smaller than expected gaps in the minefields. Darkness, dust and congestion delayed the advance. Nevertheless, by dawn on 24 October, part of one formation, the 10th Armoured Division, had reached the Miteirya Ridge. Major General Gatehouse, the divisional commander, found that his tanks were being destroyed rapidly by German artillery and requested that the attack be stopped; in a heated exchange, however, Montgomery ordered him on.[7] The next day the armour did what it could, but was unable to break open the German defences or exploit the situation. The infantry too were exhausted and, despite the 'leapfrogging' of units (or 'passage of lines'), they were unable to push any further.

Yet although the original idea of a breakthrough in one phase had not been achieved, the Allies managed to 'bite and hold' a significant section of the

German positions. Montgomery knew he had to keep the Germans off-balance, and so now orchestrated his southern feint. The attacks went in, and, as predicted, Rommel, who had just returned from convalescent leave, was unsure where the main thrust was. He had to leave his main armoured formation, the 21st Panzer Division, and the Italian Ariete Division, in the southern sector until the situation cleared.

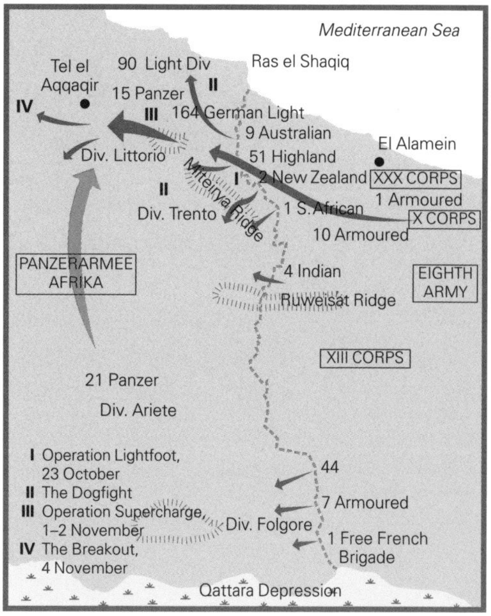

The power of the offensive, as demonstrated at the Battle of El Alamein, 23 October to 5 November 1942.

Off-Balancing and the 'Dogfight'

On 25 October, Montgomery pulled back his battered 10th Armoured and cancelled further infantry frontal attacks to facilitate an Australian thrust northwestwards against the German 164th Light Division on the coast. The change of direction made it possible to throw the Germans off-balance; the attack was covered by constant air and artillery bombardments, and there were limited infantry assaults on the main German positions in the centre. These smaller attacks were sharp and costly, but inflicted heavy losses on the Germans too – losses they could not easily replace. The Australians made a second thrust towards the coast during the night of 30 October, where they met stiffer resistance. However, they managed to sever the road and rail lines, threatening the German Light Division's line of retreat. Rommel spotted the danger and personally supervised the counter-attack. Fighting was desperate, with both sides taking very high casualties. Concentrations of German armour and Allied gunfire raked the relatively small area around Ras el Shaqiq. By the end of this phase of the battle, the Germans had lost much of their reserve armour and the bloodied Australians were still in possession of their ground.

The Final Phase: Breakthrough and Breakout

Montgomery could now resume his attack against the German centre, and Operation Supercharge was launched in the area of Tell el Aqqaqir.[8] The attack was preceded by another great artillery and air bombardment, and, again, the infantry were ordered to break into the German positions. Two British brigades made a textbook advance behind a curtain of gunfire on 2 November, with 800 guns concentrating their efforts on just 4,000 yd (3,700 m) of the German line. It was a success. Montgomery could order the long-awaited armoured thrust, but this time it was possible to deploy his formations in force.

The 9th Armoured Brigade was delayed in its pre-dawn attack, which meant that the tanks were silhouetted against the morning sky. Although they overran the German lines – and penetrated into the German artillery positions – they were clear targets for the accurate fire of 88 mm guns: 75 of the brigade's 94 vehicles were rapidly knocked out. However, much of the 1st Armoured Division had been brought up and they were ready for the German counter-attacks of the 15th and 21st Panzer Divisions, the 90th Light Division and the Italian Littorio and Trento (mechanized) divisions. The armoured action lasted all day, with vehicles weaving between the burning wrecks. Allied losses were heavy, but the German casualties were heavier still. More importantly, the Axis losses were irreplaceable. Rommel was down to 35 tanks, and he had less capability to refit and repair his damaged vehicles. By contrast, the Allies could get theirs back into action.

Rommel decided to withdraw as much of what remained of his force as he could on 3 November, and he began thinning out his troops behind an infantry and supporting gun line. Armoured cars of the 9th Royal Lancers broke through the German lines and began driving at speed into depth. The Germans were cleared from the central part of the front, and both armoured and infantry formations began to surge through the gap. Some 35,000 Axis prisoners were taken. In the pursuit, a combination of bad weather, poor communications and hesitation served to frustrate the complete envelopment of the Germans, but Rommel was unable to mount such a strong defence in North Africa again.[9] With Allied landings in Morocco and Algeria on 8 November, the Deutsche Afrika Korps was cut off.

Analysis

Several factors were critical to the outcome of this battle. Planning and preparation were thorough, from training to logistics. Recognizing and then concentrating greater numbers at the centre of gravity was crucial. As a rule, the ratio of attackers to defenders must be at least 3:1. Off-balancing attacks and multiple fronts allowed Montgomery to retain the initiative, while changes in the axes of attack drew in enemy reserves: the threat to cut off the German 164th Light Division forced the Germans to risk heavier losses to neutralize this thrust. The momentum of the attack was sustained, reaching a culminating point where the breakthrough and breakout could be achieved.[10] At El Alamein, the centre of gravity was the German army itself, not a geographical feature. Montgomery's offensive sought to reduce the capacity of the Afrika Korps to continue the campaign by inflicting heavy losses through concentrating maximum force, manpower, air power and firepower against it.

The centre of gravity is not the same in every campaign, or in every battle, and the good commander must be alert to such variations. Strict focus on the centre of gravity is also likely to be demanding: it entails the avoidance of distracting options, which may often appear easier, and the willingness to endure heavy losses. Determination, high morale and solid unit cohesion are essential. At El Alamein, the British, South African, New Zealand and Australian divisions of XXX Corps had a strong sense of *esprit de corps* and a willingness to engage in bitter fighting. Montgomery understood that, in the end, the key to this battle was the state of mind and morale of his personnel,[11] and he took every opportunity he could to instil the offensive spirit:

> Sometimes I spoke to large numbers from the bonnet of a jeep, sometimes I spoke to just a few men by the roadside or in a gun pit. I would address them less directly by means of written messages at important phases in the campaign or before a battle. These talks and messages fostered the will to win and helped weld the whole force into a fighting team which was certain of victory.[12]

2

Cambrai, 1917

Counter-Attack

The concept of the counter-attack is simple. When an attacking force has been reduced by casualties and is becoming exhausted, and its ammunition supply has been depleted, the sudden shock effect of a counter-attack can send it reeling. If the pursuit is maintained, it is possible to throw an attacker back to their start line, and even to break their morale completely. At the strategic level, Machiavelli, the Renaissance political theorist, believed that the opportunity for a counter-attack comes 'from the negligence of an enemy, who after victory, often grows careless and gives you a chance to defeat him'.[1]

Clausewitz stressed that the counter-attack gives the defender back the initiative, identifying both an inherent strength in defence and the problem the attacking force can face in terms of gathering intelligence – which tends to give rise to fatal delays or a pause in the offensive. Moreover, an attacking army can sometimes reach a 'culminating point' (see **13**), where attrition or logistical difficulties leave it vulnerable and weakened. The counter-offensive allows the defending force to dominate operations again. Clausewitz had Napoleon's Russian campaign of 1812 (see p.182) in mind when he described this phenomenon, but there are other historical examples of note.

2

Previous pages: A British tank crashes over German trenches at the Battle of Cambrai, November 1917. Insufficient reserves prevented exploitation of initial success, however, and left the British open to counter-attack.

A counter-attack can have an important psychological effect on an attacker, rendering superiority in numbers and equipment irrelevant. In the Three Kingdoms Period in China (220–80), the warlords Sun Quan of the Wu dynasty in the southeast and Cao Cao of the northern Wei dynasty struggled for control. In 217 Sun Quan had advanced to besiege the town of Hefei on the Yangtze with 100,000 men. It looked as if the Wei garrison there had little chance,[2] but their commander was the resourceful and courageous Zhang Liao. His orders from Cao Cao were to sally out and meet the enemy, despite the odds. Zhang Liao understood that a spirited and sudden counter-attack could demoralize and delay the besieging force and so, at dawn, his force of between 800 and 1,000 mounted troops dashed out of the town. They quickly broke the enemy nearest the settlement and galloped towards the main Wu camp. Panic spread, and Sun Quan had to make a personal intervention to stabilize the situation. He ordered that Zhang's force be encircled and destroyed. Zhang was undaunted, fighting his way out of the closing ring and turning back on to the Wu rear. The shock effect of Zhang's initial charge, this manoeuvre and the desperate hand-to-hand mêlée was too much for some of the Wu troops, who broke and fled. By noon, Zhang's exhausted band was compelled to withdraw, but the besiegers never recovered. Disease reduced their numbers, and the demoralized Wu army withdrew a few days later.

Cambrai, 1917

The classic counter-attack is vividly demonstrated by the German army in the closing stages of the Battle of Cambrai during the First World War (1914–18). The British had conceived of a raid by a mass of tanks on the relatively dry ground of the Cambrai-Cambresis area. With the Third Battle of Ypres still in the balance to the north, there were concerns that there were insufficient resources for another large-scale attack at Cambrai; a smaller operation meant demands for munitions and men would be limited. Nevertheless, once infantry and armour co-operation had been rehearsed, and objectives agreed, the final plan was more ambitious. It seemed possible that, committed to defending the area east of Ypres, the Germans would be unable to stem either a large-scale 'raid' or a strong advance on Cambrai.[3]

The British attack began successfully at 0620 hrs on 20 November 1917: supported by 1,000 guns delivering a short, intense bombardment on specific targets, 378 fighting tanks and 91 engineer, communications and supply tanks rolled over the German outpost line.[4] New techniques in artillery target acquisition and bridging anti-tank obstacles proved effective: successive lines of trenches were overrun on the first day, leaving a great 9-mile (14-km) wide dent, to a depth of 10,000 yd (9,000 m), in the much-vaunted Hindenburg Line defences (see p. 114), with 120 guns and 7,500 prisoners captured. But the British tanks had suffered heavy losses (179 fighting vehicles were knocked out or had become bogged down), making it difficult to maintain momentum in the days that followed. The III Corps was forced to fight hard for every inch of Bourlon Wood; the infantry, tired and depleted by casualties, now held a salient created by the attack. This left them vulnerable to bombardment from three sides. Getting ammunition and supplies across the cratered ground was difficult; the British attack was slowing to a standstill.

German Counter-Attack Doctrine

In the German Imperial Army, offensive action was the dominant philosophy. All pre-war doctrine and training had emphasized the need for bold decisions and attacking wherever possible. Remaining on the defensive was discouraged – at best a temporary arrangement until forces could be gathered to resume the offensive. If any portion of German entrenchments was captured, an immediate counter-attack was required. And if this instantaneous action was not possible, or failed, a more methodical counter-attack was to be prepared.[5] When the German offensive of 1914 on the Western Front, the famous Schlieffen Plan, failed at the Battle of the Marne, it was clear that a greater emphasis on defence was required. However, despite the elaborate and deep trench systems that evolved, the German army never lost sight of the need to counter-attack at the critical point (*Schwerpunkt*). Speed was of the essence, and quick decision-making and action were emphasized, even if the wrong decision was taken. Making faster decisions than the enemy, or *Schlagfertigkeit*, required good communications and a simplified chain of command, with delegation of decision-making to the lowest possible level. German Non-Commissioned Officers had always been given far

German 'storm troops': the vanguard of the lightning counter-attack.

more responsibility than their British or French counterparts, but, by 1917, a new tactical system had evolved that gave these men a central role in the Cambrai counter-offensive, and throughout 1918.

As early as 1914, a Guard Rifle battalion had recaptured a trench by flowing around the strongest part of the defence. One participant, Captain Willy Rohr, was subsequently tasked to create a specialist half-battalion of 'storm troops'. Sections of men, supported by their own machine gun teams, flamethrowers and light artillery, were given responsibility to break into a gap in the enemy line and then use grenades to bomb their way along a section of trench. This gave them the opportunity to attack a vulnerable flank, known as *Aufrollen*, or even to approach from the rear. The fighting at Verdun in 1916 (see p. 202) led to the formation of several larger assault battalions using the storm troops model. Even rifle battalions began to create a specialist wing of storm troops. By 1917, storm units consisted of five infantry companies, one or two machine gun detachments (each with 12 machine guns), a close support artillery unit of four to six guns, and a mortar company with eight mortars. There was also a division of roles: some units concentrated on *Stosskraft* (assault), while others provided a firebase (*Feuerkraft*) to cover the attacking troops as they closed with the enemy.

Storm troops were carefully selected and took on an ethos of *corps d'élite*. They were the fittest, keenest and most aggressive men. Ernst Jünger's famous account of the war, *Storm of Steel*, illustrates their attitude: 'bravery, fearless risking of one's own life, is always inspiring. We too found ourselves picked up by ... wild fury, and scrabbling around to grab a few grenades ... [we were soon] tearing along the line.'[6] They were the spearhead of any counter-attack, followed up by pioneer units to clear obstacles and regular rifle units deployed against any strong points that had been bypassed and surrounded. Storm troops focused on slipping between pockets of resistance, bombing to keep the defenders' heads down while they infiltrated and moved in depth.

Neither storm troops nor the regular infantry, however, could succeed without the support of artillery. In the First World War this was the supreme arm, and artillery tactics were gradually refined throughout the conflict. Lieutenant Colonel Georg Bruckmüller centralized German fire control in actions against the Russians on the Eastern Front. He championed the idea of a short but intensive bombardment which would neutralize and disrupt (but not necessarily destroy), and allow the infantry to achieve some element of surprise. A combination of smoke, gas and high-explosive shells was used to create confusion. Guns were grouped and allocated specific targets in advance. A creeping barrage (*Feuerwalze*) dropped a curtain of fire just in front of the advancing troops. A signalling system of green flares enabled the infantry to speed up the barrage if resistance was light, and field guns were manhandled forwards with the infantry to blast away at any strong points at point blank range.

The Counter-Attack at Cambrai

At Cambrai, the German plan was to counter-attack the northern and southern sides of the Allies' salient, which would force the British to abandon the central portion for fear of being cut off. There were to be two thrusts. The first would start from the south and attack northwards. Once complete, a second blow would cut from the north in a southwesterly direction.[7] Large numbers of reinforcements were brought up for the attack – 19 divisions were assembled within a week. On 30 November, a short

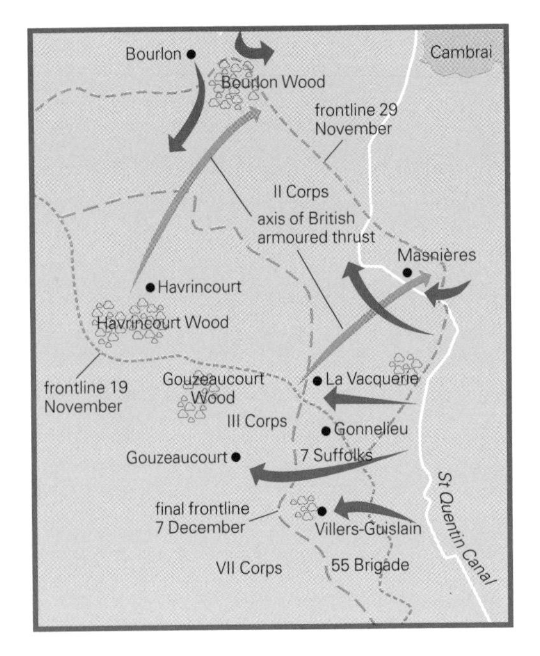

The German counter-attack at Cambrai, 30 November 1917 (shown in green), exploited British exhaustion.

bombardment followed by the swift advance of the storm troops prevented the British from having time to bring up reserves. Behind the assaulting storm troops, who swept around and between the strongest points, came the 'assault blocks' of battalion strength, with mortars, machine guns and infantry. Their task was to maintain the momentum of the attack. Further back came the pioneers and then the regular infantry who could mop up, repair bridges and remove obstacles. German aircraft strafed the British positions just ahead of their advancing infantry.

Despite the speed and weight of the attack and the use of thousands of gas shells, the Germans made little progress at Bourlon Wood in the north. The same difficulties that had attended the British attack on this feature also confounded the Germans, particularly the broken nature of the ground. But there was another factor. The British may have been tired and depleted, but even isolated groups offered stiffer resistance than expected. German casualties were correspondingly high.

In the south, the Germans managed to break in between the British divisions at Villers-Guislain. The 7th Suffolks, attacked from all sides at Gonnelieu, were wiped out. There was confusion across the British sector, but the arrival of the 1st Guards Brigade, who took back Gouzeaucourt, stabilized the situation. In the centre, the survivors of the defence held out in La Vacquerie. Nevertheless, the German counter-attack continued, and in the southern sector the Germans not only regained their original positions, but actually pushed the British back beyond them. In the north, the British managed to hold on to most of their gains, relinquishing ground to a depth of 2 miles (3 km).

The price of this battle for both forces was heavy: in the region of 40,000 dead on each side. However, the German counter-attack had restored their line, and their techniques might have achieved even greater success had it not been for the extraordinary resistance of the British units engaged and the timely arrival of reinforcements.

Analysis

There were several reasons for the German success at Cambrai. The German army struck once the British offensive was exhausted and at its culminating point, as was clear from the limited scope of fresh attacks, the inability to seize more ground and the general slackening in the tempo of operations. Another factor was the Germans' ability to co-ordinate all arms and to concentrate their firepower at precise locations. They used deception and achieved surprise (keeping tight security on their build up, and using the shock of a sudden bombardment). They also managed to obtain a local superiority in numbers. Most important of all, they achieved great speed, in part by devolving command to more junior levels. Storm troop commanders were expected to use their initiative, seek out gaps and exploit them. The speed and shock effect of their counter-attack denied the British the chance to reinforce their lines or plug the gaps. It also had a psychological effect. But for the tenacity of the defenders and the arrival of fresh Allied reinforcements, the Germans might have been able to achieve even more.

C. E. Callwell, writing in the 1890s, believed that every opportunity should be sought to keep a defence 'active'. He wrote: 'a great chance often presents itself of seizing upon the moment when the enemy is disordered by [an] advance, to deliver a crushing counter-attack'[8] and conlcuded that 'Resolute initiative is the secret of success.'[9] In another era, General George S. Patton stated 'In war the only sure defence is offence, and the efficiency of the offence depends on the warlike souls of those conducting it.'[10] The German counter-attack at Cambrai illustrates the effectiveness of this analysis.[11]

3

Teutoburg Forest, AD *9*
The Six Day War, 1967

Surprise Attack and Ambush

The initial strike in any campaign will always confer an advantage. An element of surprise can augment the effect, and an unprepared enemy will suffer a disproportionate physical and psychological blow. Throughout history the surprise attack has been a popular 'force multiplier'. At the lowest tactical level the ambush at close quarters – sometimes orchestrated by just a handful of men – can have devastating results. At the grand strategic level the outcome is often no less dramatic. On a number of occasions a numerically inferior force has used a preventive strike or an ambush to defeat a larger and apparently stronger enemy.

The principles of an ambush or a preventive strike are similar. In choosing the site of a surprise attack, good intelligence is vital; reconnaissance, detailed planning of every aspect of the operation and rehearsal help to capitalize on this. To maintain the element of surprise, tight security is clearly essential, and this extends to troops and their civilian contacts or logistical sections. Concealment and deception can assist in this regard. To avoid confusion and to ensure that the trap is sprung at precisely the right moment, clear communications and a simple plan are also required. At the point of the attack, the force deployed must be overwhelming, backed up by high-quality training to ensure battlefield discipline and alertness.

3

Previous pages: Israeli troops, here advancing in convoy, achieved a stunning success in 1967 when they took Egyptian forces by surprise in the Sinai.

Teutoburg Forest, AD 9

In AD 9, Varus, the Roman governor of northern Germania, received reports of unrest along the Rhine. The region had been incorporated into the Roman empire for a relatively short time, and the emperor, Augustus, intended to extend his dominion eastwards and bring the Germanic tribes more firmly under his control. The Roman historian Tacitus described the Germans as fierce fighters, but also noted that they lacked organization. Clans and families were gathered into *ad hoc* units, and few were armed with anything more than light spears.[1] Mounted men carried swords and shields and fought among the foot soldiers. Tactics consisted of hurling spears, charging in small wedges and then retiring to recover before attacking again.

Intelligence about the rebellion was provided by Arminius, a Romanized German and officer of a Roman auxiliary cavalry unit. However, what Varus did not realize when he despatched three legions (XVII, XVIII, XIX) through the Teutoburg Forest was that Arminius was actually the leader of the revolt and that the Romans were marching straight into an ambush. The site of the attack was carefully selected. The 20,000 Romans were strung out along the line of march with a lengthy baggage train behind, while the forested and swampy terrain they were traversing made it almost impossible to deploy in the classic Roman formations.

A Roman face helmet found at Kalkriese Hill, identified as the site of the ambush of Roman legions by Germanic warriors.

The first attacks used the element of surprise. Some accounts suggest the Roman cavalry was separated from the infantry, while others hint that the Roman foot soldiers were themselves isolated in smaller detachments where they were destroyed in detail by the more numerous German warriors.[2] Despite their advantages, it took three days of sustained fighting before the tribesmen could overcome the legionaries. The Roman historian Dio wrote that a storm had thrown down boughs and made any kind of order impossible, but that after the first attacks the Romans built a fortified camp.[3] An attempt was then made to continue the march. As they entered more forested terrain, and in torrential rain, the Germans attacked again and the Romans 'suffered their heaviest losses'.[4]

Although much of the detail of these actions has been lost, recent archaeological discoveries suggest that the Romans may have made a final stand at Kalkriese Hill, north of present-day Osnabrück. The Romans were massacred; there were almost no survivors. Varus committed suicide and his head became Arminius' trophy. Through effective exploitation of surprise and terrain the German tribes, in spite of their limited central organization, had annihilated a substantial field army deployed by the largest contemporary military state. This psychological victory ensured that Germany remained free from Roman control, with the Rhine established as the long-term frontier.

The Six Day War, 1967

Another classic example of surprise attack, on an even greater scale, occurred during the Arab-Israeli Six Day War. In 1967, the Israelis were alerted to the belligerent intentions of the surrounding Arab states by a combination of troop movements towards their borders and increasingly warlike rhetoric on Arab radio stations. The increase of Syrian air raids and shelling from the Golan Heights (which overlook northern Israel), indicated that attack was imminent, while Cairo media stations claimed that Israel was about to be 'wiped off the map'.[5]

Over several weeks the Israeli Defence Force (IDF), a significant proportion of which was made up of citizen soldiers, was quietly mobilized and a preventive strike planned: using the element of surprise, there was to be an overwhelming attack. The first strike was vital because Arab forces outnumbered the Israelis; moreover, the Egyptians had been equipped with the latest Soviet armour and aircraft. Nevertheless, quality of weapons is never a guarantee of success – it is the men and women who operate them, and their level of training, experience and determination, that really count. In the Six Day War, devolved leadership was also decisive.

Air Operations

At 0745 hrs on 5 June 1967, the Israeli Air Force (IAF) made its first attacks against Egyptian air bases, flying low to come in under Egyptian radar. Egyptian aircrews were still off their bases at this hour, while their aircraft were neatly laid out rather than being dispersed, factors which assisted the assault. In the first three hours, the IAF flew 500 sorties against 19 Egyptian bases in the Sinai, Nile Delta and Cairo area, destroying 309 out of 340 serviceable combat aircraft, including bombers, fighter bombers, fighters, transport

aircraft and helicopters.[6] The destruction was so complete that the Israelis enjoyed almost complete air superiority for subsequent ground operations in Sinai. The other Arab states suffered similar treatment. Communications systems and anti-aircraft batteries were taken by surprise, making countermeasures impossible. By the end of this first day, the Jordanian air force had ceased to exist, with all its aircraft destroyed. The Syrians lost 55 fighters and 2 bombers, two-thirds of their strength. By the end of the second day, the Arab air forces had lost 416 aircraft, the majority of which had never even got airborne. The Israelis lost only 26 planes.[7]

Opposite above: Israel's air and land campaign took the Egyptian forces by surprise with devastating results.

Opposite below: Wreckage of an Egyptian plane; much of the Arab air forces were destroyed on the ground, leaving the Israelis with almost complete air superiority.

Ground Operations

The Israelis knew that air attack, while effective, would not be enough to forestall an invasion – a land operation would also be needed. Several critical points also made remaining in a defensive posture impossible. First, the numerically superior Egyptian forces were poised to strike from the Sinai and Gaza. Attacking them would be exceptionally difficult, however, as the entire eastern Sinai comprised a series of interlocking bases and field obstacles, including minefields and concrete strong points. Secondly, the Israelis faced enemies on all fronts and would have to cover every mile of their eastern border against Jordanian, Iraqi and Syrian forces.

The Syrians held the strategically important Golan Heights, from where they could observe Israeli movements in the plains below. If the Israelis waited to counter Arab attacks, there would be heavy civilian casualties and they would be overwhelmed since they could not defend all sectors strongly. The plan therefore was to strike simultaneously against the centre of gravity of the Egyptian, Jordanian and the Syrian forces, the three largest threats. In the Sinai this meant a three-pronged armoured thrust, while in the Golan and the West Bank, carefully synchronized air, armoured and infantry assaults would be needed.

In the Sinai, the Egyptians had five infantry divisions and two armoured divisions, totalling 100,000 men and about 1,000 tanks, with generous artillery support. The infantry were deployed evenly on the most likely axes of approach, with the armour held back in depth. Against them, Israel could muster three armoured divisions. To even the odds, the Israelis had carried out a deception plan prior to the opening of the war, in which their own armoured units appeared to be heading south along the border to reinforce the area around Eilat at the head of the Red Sea. However, at 0800 hrs on 5 June, the three Israeli armoured divisions attacked across the northern Sinai border, once again achieving surprise. To avoid the minefields south of Rafah, the first division, under the innovative Major General Israel Tal, drove straight along the highway from Khan Yunis, fighting its way through the town and on towards El Arish.[8] This enabled the division to avoid the successive lines of concreted emplacements and deal with the 100 tanks held in reserve west of Rafah. At the same time, a parachute brigade and a battalion of tanks made a wide sweep to the south, getting behind the defences at Rafah and attacking the Egyptian artillery lines to take successive positions in textbook style. This same brigade then again achieved surprise on the second Egyptian line by drawing out their tanks with a feint, before taking them in the flank with a concealed armoured group. Pressing on towards El Arish, they overwhelmed the shocked defenders. The Egyptians rallied briefly and retook the position, forcing Tal to fight through a second time, but by the end of the first day, Tal's armour was in possession of the town.

The second armoured division, led by Major General 'Arik' Sharon, pushed southwestwards on the Nitzana-Ismailia axis, overwhelming a well-entrenched Egyptian infantry division. Once the outpost line was driven in, he used the divisional artillery to neutralize the main defences. Next, he ordered flanking units to extend on either side and to reach into depth, where they could provide intelligence on the movements of the Egyptians and intercept any reinforcements. The northern wing faced some tough fighting, but was soon positioned behind the Egyptian defences. A parachute unit was then flown in by helicopter to destroy the Egyptian artillery. With the main position isolated, Sharon's battle tanks rolled on to the defences, supported by infantry clearing the trenches. By 0600 hrs on 6 June, the Israelis had overrun the entire complex.

In the centre, General Yoffe's division crossed the open country between Sharon's and Tal's axes. The Egyptians had left the area undefended, assuming that the sand dunes were impassable. Yoffe's objective was to prevent lateral movement between the two axes of advance and to intercept reinforcements. As Sharon's division was battling away to the south, Yoffe's forces sped forwards and reached the crossroads at Bir Lahfan. Hull down (dug in) and dispersed, they surprised an Egyptian armoured brigade making its way south at nightfall and destroyed it in a matter of hours.

By the end of the second day, the Egyptian bases around Jebel Libni had been captured and everywhere the Israelis were still advancing westwards. By contrast, the Egyptians were in confusion and units began to withdraw towards the Mitla Pass, a choke point on the route to the Suez Canal and their next defendable line. The Israeli Air Force was free to strafe retreating Egyptian units and Israeli ground forces drove past columns of burning vehicles.

Fast-moving Israeli armoured and mechanized forces dash across Sinai in 1967.

The Mitla Ambush

The Egyptian withdrawal to Mitla presented a golden opportunity to inflict a decisive defeat on the remaining two Egyptian armoured divisions in that area. While Sharon chased from the east, the other Israeli divisions began to converge. Racing ahead, Colonel Yiska Shadmi of Yoffe's division planned to block the eastern entrance to the Mitla Pass. Operating at their maximum range, four of the eight tanks available ran out of fuel and had to be towed to the summit. There was barely time to dig in before the leading Egyptian elements began to make their way up the pass from the west. This hasty ambush took the Egyptians by surprise and the results were devastating: the little Israeli force destroyed several vehicles, even while having to scavenge fuel and ammunition from destroyed Egyptian tanks until the rest of Yoffe's force arrived. To the east, the other Israeli armour, including Sharon's division, was fast approaching. The Egyptians made desperate efforts to escape, but each attack failed to break through.

As in many conflicts, this dash produced a bizarre vignette. On the night of 7 June, an Israeli unit had become mixed up with a retreating Egyptian road column. The Egyptians seemed completely unaware of the Israeli tanks in their midst and took no action. However, after proceeding for some distance, the Israelis turned their tanks off the road on a single command, drove into line, threw on their searchlights and opened fire. A perfect ambush, efficiently executed: the entire Egyptian force was destroyed.[9] The Israelis soon reached the Suez Canal. Behind them the Egyptian army lay in ruins.[10] About 80 per cent of their equipment was lost in the rout: 15,000 men, 800 tanks, 10,000 vehicles, several hundred guns, plus 5,500 prisoners. It is estimated the Israelis lost 300 men and 60 tanks.

Analysis

Surprise enabled the German tribes, whose relative disorganization was despised by the Romans, to eliminate an entire, large Roman field army. Quite apart from the loss of trained men, the shock of the disaster convinced Emperor Augustus that it was not worth persisting with the attempt to annex territory east of the Rhine. No subsequent emperor dared to reopen this project.

Similarly, in 1967, the element of surprise allowed three Israeli armoured divisions to capture the entire Sinai peninsula in just three days, against substantial odds.[11] An overwhelming first strike by the IAF knocked out the Arab air forces on the ground. Egyptian units could now be attacked from the air, and Egyptian command and control mechanisms were severely disrupted, making it difficult for them to regain the initiative. On the ground, outstanding leadership and initiative at every level of command maximized the Israeli opportunities and the Egyptians were repeatedly surprised by armoured ambushes. The Israeli strategy was aggressive and effective.

The Arab disaster was also partly of their own making. They had been preparing for an offensive but security was poor and the most obvious precautions were not taken. The Egyptians had also anticipated a long drawn-out campaign, so their initial response to the Israeli attack was slow. Egyptian propaganda had gone into overdrive, claiming that the Israelis were being beaten on their own territory, but this simply misled Egypt's allies and made the defeat even more shocking.[12] Egypt recovered quickly and learnt the lessons so painfully taught in this war. In 1973, it was they who managed to surprise the Israelis (see p.171), but a variety of factors conspired to deprive them of a victory on the scale of 1967.

4

Walaja, 633
Operation Uranus, 1943

Envelopment and Double-Envelopment

The unexpected appearance of enemy troops on a flank or from behind can have a damaging effect on an army's morale. If a force is encircled it can be deprived of supplies or attacked from any quarter. Ultimately, if completely cut off, it must either thrust its way out, surrender or fight to the death. Envelopment is the classic example of manoeuvre warfare, and the double-envelopment, or encirclement, has produced some of the most famous and decisive victories in world history.

To be successful, the envelopment must fulfil a number of requirements – mobility and speed are essential, and the enemy must be pinned, or at least unable temporarily to break out or manoeuvre. In many examples, deception and good security have been critical in order to prevent enemy counter-measures.

An example from the Roman Republic clearly illustrates these principles in action. At Cannae (216 BC), during the Second Punic War, the Roman Consuls Varro and Paullus were defeated by Hannibal's Carthaginians in one of the most celebrated double-envelopment battles.[1] Hannibal had positioned his weaker infantry in the centre of his line, and this force retreated as the Roman infantry advanced. The Carthaginian withdrawal was accompanied by the extension of cavalry and light infantry on both flanks, which eventually enveloped the Roman troops. Unable to manoeuvre and hemmed in by the Carthaginians on every side, the Romans suffered catastrophic casualties.

Previous pages: Red Army troops storming an apartment block in the ruins of war-torn Stalingrad. The endurance of Soviet soldiers and the sudden appearance of fresh troops in late 1942 baffled the Germans.

In Asia in the 13th century Genghis Khan also understood the effectiveness of envelopment.[2] Mongol horsemen adapted a technique from hunting for use in battle: riding hard in a concentric pattern around a grazing area in two wings, a hunting party would gradually close inwards towards its prey until archery could be brought to bear in a relatively confined killing area. In the campaign against Bukhara in 1220, Genghis Khan used the convergence of his armies in the same way.

Walaja, 633

Not all envelopments or flanking attacks occurred in the manner of Cannae or Bukhara, and an interesting example can be found in the Islamic conquests of Khalid ibn al-Walid. In 633 the early Islamic Caliphate attacked the Sasanian Persians, inflicting consecutive defeats at the Battle of the Chains and the Battle of the River in Mesopotamia. This opened the way to Hira, the regional capital (in present-day Iraq), but the Muslim commander, Khalid ibn al-Walid, had to confront two converging Persian armies.[3]

The first Persian army, commanded by Andarzaghar, was sent to check the Muslims at Walaja, a key location for Khalid if he had to withdraw back into the desert. Survivors of the previous battles and several thousand Arabs joined the Persian army as it progressed through the region. With more troops on the way from Ctesiphon, the Persians felt confident that their superior numbers (now about 30,000) would decide the action.

Khalid was fully aware of the Persian build up at Walaja since he was receiving regular updates on Persian movements and strengths from spies among the disaffected Arabs of the Tigris. He also knew that his 15,000 men were confident of defeating the Persians, as their recent battle experiences had demonstrated, while the prospect of rich spoils from the wealthy Persian empire inspired them. Khalid decided that he had to attack before the two Persian armies combined, and, given that so many Persians had regrouped after the Battle of the River, he wanted to inflict the heaviest casualties possible. To effect this, he envisioned a double-envelopment manoeuvre. First, he posted a number of piquets along the lower Tigris to guard against a surprise enemy crossing, while he moved his army rapidly to reach the Persian camp at Walaja.

Khalid used the configuration of the battlefield to his advantage. The Persians were deployed on a 2-mile (3-km) wide plain between two low ridges. The night before the action, Khalid sent 2,000 of his cavalry in a wide flanking march and ordered them to remain concealed behind the western ridge, beyond the Persian right flank. In order to maintain tight security, the majority of the Muslim army was unaware of this deployment. The plan was for the rest of the army to pin the Persians until the critical moment. On their part, the Persians expected the Muslims to make a wild charge and then, once this attack had exhausted itself, they would counter-attack. As a result, the Persians' posture was initially defensive.

On the day of the battle, Khalid ordered his infantry forward, and, as expected, the fighting seemed to go in favour of the more numerous Persians, who fed reserves into the battle to replace casualties. As the Muslim troops tired,

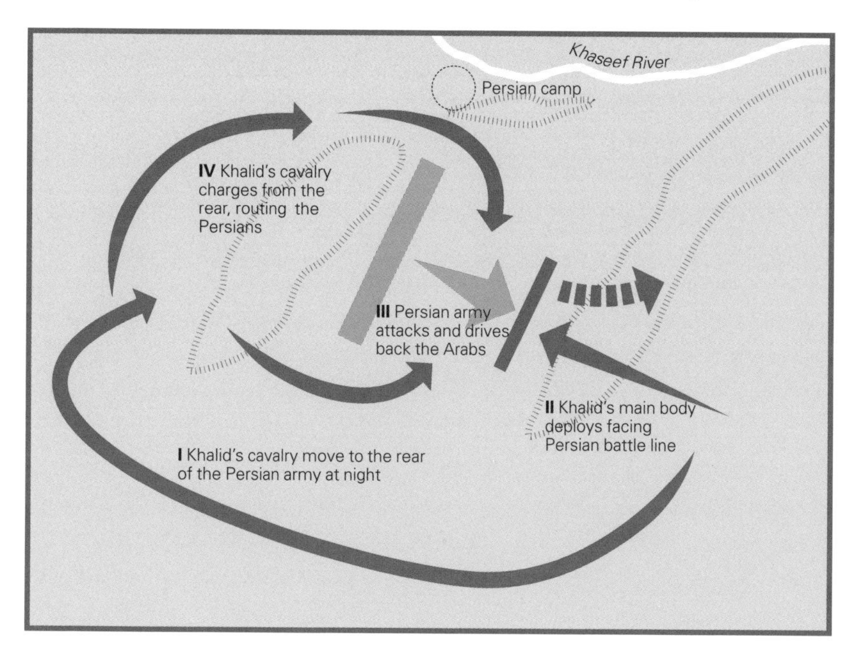

Khalid ibn al-Walid's stunning victory over the Persians at Walaja in 633 was achieved by the use of patient and calculated envelopment.

Andarzaghar ordered his men to counter-attack. For a time, the Muslims appeared to be able to hold their line, but as they were pressed back, Khalid gave the order for the hidden cavalry to be unleashed. This sudden charge from the rear caused shock and confusion among the Persians, and re-energized the Muslim main body, which resumed its attack. The Persian army was almost surrounded and found it impossible to manoeuvre. Its numerical superiority was negated as the centre was unable to bring its weight to bear and the demoralized troops looked to escape. The relatively small number of horsemen involved was clearly not appreciated in the heat of the battle, but the shock effect of the fresh Muslim force from an unexpected direction threw the Persians off-balance. Eventually the Persian army collapsed and fled.

The consequences of the battle were significant. The Muslims, although depleted and exhausted, went on to defeat another Persian army at Hira and eventually captured Iraq. Although the Persians recovered and retook the region, the Muslim armies had broken the psychological superiority of the Sasanian Persians. In 636, the empire was invaded for a second time and fell under Islamic control.

Operation Uranus, 1943

One of the clearest examples of a decisive victory brought about by a successful envelopment manoeuvre can be found in Operation Uranus, the Soviet counter-attack against the Germans at Stalingrad in the Second World War.

The German invasion of the Soviet Union in 1941 had driven some 550 miles (900 km) eastwards and come to within striking distance of Moscow when winter conditions slowed the offensive. In the spring of 1942, Hitler had been advised to avoid a direct assault on Moscow and to concentrate on other fronts where greater strategic gains could be made. In the southern sector, the fall of the Ukraine held out the possibility of advancing further eastwards in good tank country, perhaps as far as the oil-rich lands of the Caucasus. There was also the possibility that the supply artery along the Volga River could be denied to the Soviets, rendering the defence of Moscow more difficult. Furthermore, Stalingrad was a production centre for tanks and munitions; its obvious ideological and propaganda symbolism enhanced its significance as a prize.[4] These considerations aside, the strategic imperative was to weaken Soviet forces before the Americans could threaten the German position in Western Europe. However, the Germans lost sight of the true centre of gravity, and allowed themselves to regard the city of Stalingrad, not the Soviet armies, as their focus of operations.

Hitler's interventions during the operations caused confusion and delay.[5] First, it was decided that the southern front would be divided between two army groups, A and B. The Caucasus thrust, launched by Army Group A, was delayed, since Hitler's insistence on seizing Stalingrad in July 1942 deprived this force of vital resources. Worse, the focus on Stalingrad played to the strengths of the less mobile Soviet forces: in the street fighting that developed, German superiority in armour and air power was negated. With the German spearhead of Army Group B, the VI Army, pinned down in Stalingrad, the Soviet commanders Georgy Zhukov and Aleksandr Vasilevsky prepared a double-envelopment counter-attack.

The Battle of Stalingrad drew in thousands of Soviet troops, but Zhukov and Vasilevsky were able to build up and hold in reserve five new tank armies. To gain battle experience these were given limited objectives on other fronts, while for five months the rest of the army endured the fiercest fighting of the Eastern Front. Strict security was maintained and even the information the German high command did receive was discounted as exaggerated since the Nazis did not believe that the Soviets were capable of any large-scale mobile operations on this front.

German commanders were focused on the fighting in Stalingrad itself, where even individual buildings had become tactically significant. Losses were heavy, with 20,000 casualties each week at the height of the fighting. However, the situation on the German flanks should have been a cause for concern. In the south, the German armoured thrust was slowing down because of overstretched supply lines; their mobility was in jeopardy. To the north of

Stalingrad, less well-equipped and supplied Romanian divisions were spread too thinly along the Don to offer serious resistance if attacked in force. This fact had not escaped the attention of the Soviet planners.

At 0730 hrs on 19 November 1942, Operation Uranus began with the Soviets delivering a heavy artillery bombardment from 3,500 guns for an hour and a half, both on the Romanian and Italian positions on the Don and on the Romanian Fourth Army defences south of the city.[6] They then unleashed their armoured formations, including three tank corps. To the north, the Romanians initially put up a stubborn resistance, but their position was hopeless and they were simply overwhelmed on the first day. A pocket of Romanian forces eventually surrendered at Raspopinskaya on 23 November.

The situation was similar in the south. A combination of artillery, mortars and Katyusha rocket batteries suppressed the Romanian Fourth Army

Soviet infantry, with artillery acting in close support, burst through the flanks of the Stalingrad salient.

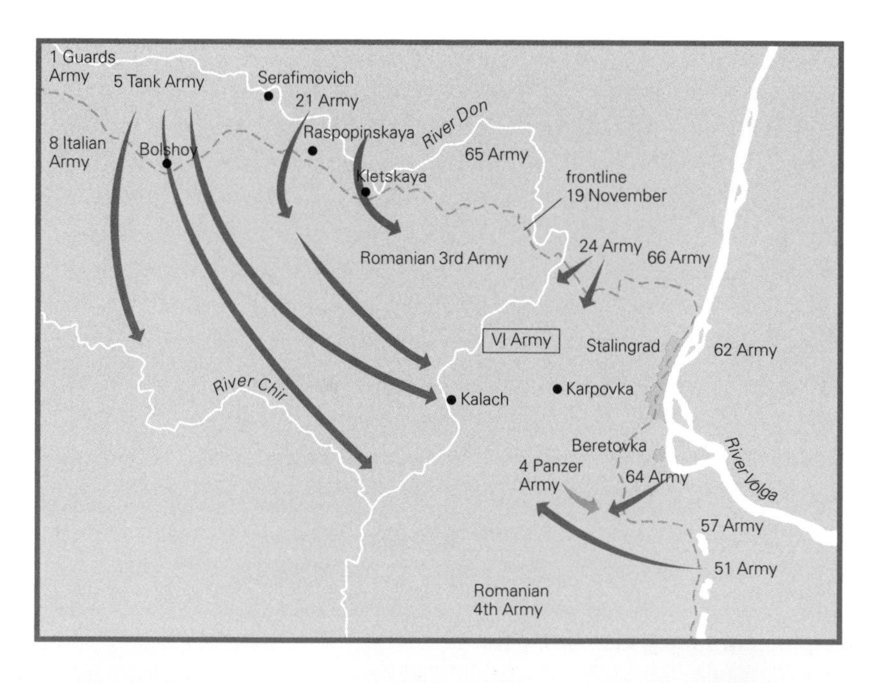

Operation Uranus, the classic envelopment, sealed the fate of the German VI Army under von Paulus at Stalingrad.

positions and neutralized their command and control systems. Two Soviet armies, the 64th and 57th, with six infantry divisions, advanced northwest behind the German forces in Stalingrad. The 13th Mechanized Corps drove towards Karpovka, the 4th Cavalry Corps swung southwest, while the 51st Army and the 4th Mechanized Corps pressed on towards Kalach, to the west of the city. The speed of the Soviet advance took the Germans by surprise. Their appearance in depth was a shock to rear echelon units, but the lack of any significant German reserves was a fatal oversight.

In just four days, the German VI Army had been completely enveloped. Attempts to cut a way through from the west by General Manstein in mid-December were initially successful, but Soviet reserves were moved in to plug the gap. The Soviets avoided attempting to reduce the Germans' perimeter at Stalingrad and instead concentrated on advancing further west. Hitler still clung to the belief that the German position in Stalingrad was strong and he refused to allow the commander there, General Friedrich von Paulus, to break out.[7] An attempted 'air bridge' of supplies failed to meet the needs of the beleaguered army.[8] Barely a tenth of the requirements made it to the garrison. Paulus's men were starved, bombarded and assaulted on all sides. The Soviets then brought up seven armies to begin the attack to retake the city. The fighting was just as intense as in the previous autumn. Once again, individual buildings, now in ruins, became the scene of bitter struggles. The Soviets had expected their final attack to take four days, but, even though the Germans had been cut off for two months, the battle went on for two weeks. Finally, crippled by the freezing conditions, dwindling ammunition and scarce supplies, Paulus could offer no further resistance. On 2 February 1943 he surrendered his forces, 90,000 strong, to the Soviets, a humiliation for Hitler and a turning point on the Eastern Front.

Analysis

In these two examples, the attacking forces possessed greater mobility and speed than their opponents. At Walaja, Khalid used his mounted arm, pre-positioning it as his striking force to the rear-flank of his enemy. The Soviets at Stalingrad also used their mobile tank armies, and even cavalry, to strike rapidly in depth behind the German forces. In both examples, the enemy was pinned by fighting in the centre. The Persians had not maintained a sufficiently strong reserve to deal with an attack from a new direction, and the troops who were already engaged did not have time to extract themselves, reorganize and manoeuvre. The Germans' focus on Stalingrad pinned them into a position, and also drew in their best units, leaving relatively weak formations to protect their vulnerable flanks. The Soviet armies enjoyed local superiority both in numbers and also in tanks, artillery and equipment, and the speed of their advance left little time for the German reserves to react.

Deception and good security were also paramount in both examples. Khalid issued clear and straightforward plans to his subordinate cavalry commanders and ensured a simple system of communication to initiate the envelopment. He had taken the precaution of not informing the rest of the army in case information was passed to the enemy. At Stalingrad, Soviet security had also been good, with strict radio silence and most movements taking place at night, but it had not been possible to conceal the build up completely. The German VI Army had warned Hitler of the threat, but he and his senior officers had refused to give credence to the information, preferring to believe that the Soviets were constructing defensive positions rather than preparing an attack. Paulus had even suggested a withdrawal from Stalingrad to counter the possibility of an attack on his flanks, but Hitler had rejected the idea, as he refused also to allow Paulus to break out of Stalingrad once he was surrounded, giving the Soviets an opportunity to consolidate their encirclement.

5

Bouvines, 1214
Chancellorsville, 1863

Flanking

Outflanking – the turning of a flank, an attack on the rear or the total encirclement of a force – offers the opportunity to neutralize the strongest part of an enemy force, namely its centre or front. By appearing from an unexpected direction or simply overlapping the enemy's flanks to cause panic, an attacking army can also inflict a psychological blow on an enemy. If an enemy is simultaneously engaged from the front, flanking also offers the chance to bring more fire or weapons to bear.

In 326 BC in the Punjab, Alexander the Great's force of at least 50,000 men was confronted on the opposite bank of the fast-flowing River Hydaspes (Jhelum) by the local king, Poros (Pururava).[1] Alexander knew that a direct frontal assault would be too costly, so he identified a second crossing point 17 miles (27 km) upstream. Crossing the flooded river at night with 10,000 troops, he defeated a small detachment sent to intercept him and pressed on to confront the main body. In the battle that followed, Alexander used his cavalry first to weaken a flank and then to encircle the Indian force, while pinning the centre with an infantry assault.

Flanking attacks can also retrieve the situation when it appears a battle is lost. At the Battle of Stilo, or Cap Colonna (982), in Calabria, the Imperial heavy cavalry of the Holy Roman Empire destroyed the centre of the Muslim line and penetrated the heart of their position. But a flanking attack by Muslim light cavalry enveloped Emperor Otto II's heavy horsemen and inflicted losses of some 4,000 men, turning the battle completely.

Previous pages: Overrun Confederate trenches at Marye's Heights, following the battle of 3 May, 1873. Despite Union success here, it came too late – Lee's outflanking operation at Chancellorsville further west compelled the Union army to abandon its offensive.

Amphibious operations have at times made it possible to effect a flanking manoeuvre to circumvent armies and stronger defences. In May 1583, the Spaniards made the 'Terceras' landings in the Azores against a combined English, French and Portuguese garrison. After a thorough reconnaissance, a diversionary landing was made to draw the defenders to a particular beachhead; the main landings then got under way on the flank of this position. On a larger scale, in the Second World War the Normandy landings (1944) and the amphibious operations at Tarawa and Okinawa (1945) were frontal attacks at the tactical level, but also opened up new fronts on the flanks of the Axis powers at the strategic level.

Classic examples of the flanking attack utilizing firepower can be found in the Franco-German War of 1870–71. At Wörth in August 1870, the Prussian general Moltke the Elder pushed his infantry around the flanks of the French forces, using artillery and rifle fire to apply pressure.[2] At Sedan, in September, he again used firepower to reach around the flank of a French army that had become pinned to the defence of the town. German infantry and artillery were extended to the high ground south of Sedan and eventually reached around the French rear.

A flanking manoeuvre need not be in the immediate vicinity of a battle to create a shift in the strategic balance. During the India-Pakistan War of 1971, the fighting along the border between armoured and mechanized formations had proved inconclusive. However, an Indian thrust across the relatively undefended Rajasthan desert towards the Indus, far to the south of the main theatre of operations, while a minor element of a larger campaign, compelled Pakistan to abandon its planned armoured counter-stroke from Lahore.[3] The Pakistan army could not be sure that India would not penetrate more deeply into the country from the south, and it could not commit its final strategic reserve under these conditions. Satisfied that it had brought the Indian forces to a standstill elsewhere, it called for peace negotiations. With a relatively small force India had, by a bold thrust on a flank, effectively neutralized the strategic value of Pakistan's armoured forces.

Bouvines, 1214

King John of England and the Holy Roman Emperor, Otto IV, backed by some French noblemen, notably Ferrand Count of Flanders, Renaud Count of Boulogne, Henry Duke of Brabant and Hugh of Boves, aimed to defeat their arch enemy, Philip Augustus of France, after a conflict lasting over a decade. King John conceived a grand flanking strategy to bring down Philip. He would attack from southwest France, forcing Philip to field an army against him, while Otto IV would thrust against Philip in the north. John's expedition, while not very successful, did tie down 600 knights and thousands of foot soldiers under Philip's son, Prince Louis, in the Loire Valley. Philip himself went north to face the coalition army and encountered them at Bouvines.

Philip had advanced to Tournai, intending to ravage the county of Hainaut, but he then discovered that the allied army was at Mortagne, just 7 miles (12 km) to the south. He decided to withdraw to Lille, some 18 miles (30 km) to the west. Some of his troops, led by a Hospitaller, Guérin, bishop-elect of Senlis, and the Viscount of Melun, apparently on their own initiative, moved south to watch the enemy. Otto IV decided to attack the French on their flank, by moving northwestwards from Mortagne to meet the Tournai–Lille road at the bridge of Bouvines. This involved taking a narrow route through dense woodland. The French deployed crossbowmen across the woodland road in an effort to delay the Germans in their march.

The French army was at that moment straddled across the River Marque at Bouvines. Philip realized that his rearguard was in danger of being cut to pieces by the advancing allies and he therefore determined to fight. He ordered his troops back over the river, building a temporary bridge of logs to speed the process. Nevertheless, it took over an hour

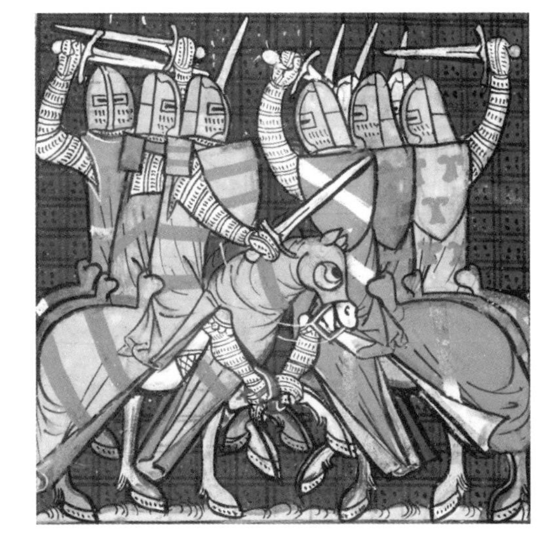

This depiction of the Battle of Bouvines in a manuscript illumination from the *Grandes Chroniques de France* reflects the critical role played by the disciplined French cavalry.

to assemble the troops into a fighting line. Fortunately for Philip, the Germans were also having trouble assembling their men, as the narrow road emerging from the forest delayed their deployment.

To gain time to get his troops back across the river, Philip acted boldly, leading his cavalry in a great gallop that forced the allies to disentangle their own cavalry and deploy them in a great mass to the south of the Lille–Tournai road. Philip then handed over command of the cavalry to Guérin, who proved to be a skilful leader. Guérin extended his front so that he could anchor the French southern (right) flank on the marsh, in the meantime sending forward light cavalry to delay the gathering mass of enemy knights under Ferrand of Flanders and prevent them from interfering with his deployment. Despite suffering terrible casualties, the light cavalry caused disorder in the enemy ranks, and Guérin then began to send squadrons – up to 150 at a time – of heavily armed knights against them. A battering duel began between the two opposing cavalry forces on the French right.

Otto IV now ordered a massive but rather disorderly frontal attack against the French line. In the confused fight for the centre, Philip was pulled from his horse and only rescued by his bodyguard. At the point when the momentum of the allied attack was spent and the battle among the footsoldiers was settling down to a grim hacking match, well-controlled French cavalry charges on the right flank succeeded in breaking the back of the allied forces and captured Ferrand of Flanders. The critical moment arrived when the allied infantry in the centre found their right flank exposed as the Duke of Brabant, who had taken no part in the fighting, fled with his forces. This caused general panic among the allied foot. French cavalry then charged into the allied centre from the flank. In the mêlée, Otto narrowly escaped with his life. As the allied army dissolved, one of its leaders, Hugh of Boves, formed his infantry into a circle and, using their long spears, held off the French. A small cavalry force sallied out to attack the French as they turned away from this bristling circle. This desperate counter-attack failed and the French closed in on both flanks.

The Battle of Bouvines changed the shape of Europe. It created a French hegemony that would last until the 'Hundred Years War', a century later, and also meant that Frederick of Hohenstaufen and Pope Innocent III soon triumphed in the war for the German Empire over Otto IV. Moreover, in England, the barons, sensing their defeated monarch's weakness, rebelled against King John, forcing him to sign the Magna Carta, the foundation of English liberties.

Chancellorsville, 1863

Major General Joseph Hooker, the commander of the Union army around Fredericksburg in the American Civil War (1861–65), had been unable to break the Confederates with frontal attacks. He therefore left a covering force opposite the town and marched west along the Rappahannock River, which he crossed on 30 April 1863.[4] His plan was to make his way through the forested region known as the Wilderness and then emerge behind the Confederates. That night, the Union forces camped around the house and estate at Chancellorsville, intending to continue their advance the next day.

Although heavily outnumbered, General Robert E. Lee left 10,000 Confederate troops to hold Fredericksburg and marched boldly to intercept Hooker's forces. At Zoan Church on 1 May, the Confederates skirmished with the leading elements of the Union army and, much to everyone's surprise, Hooker ordered a withdrawal to a perimeter around Chancellorsville. As his men threw up hasty defensive lines of logs and earth, Hooker calculated the Confederates would be cut down by concentrated fire from these defences.

General 'Stonewall' Jackson, who led his men in a bold flanking march around Hooker's right.

Lee then conceived a brilliant plan. He ordered General 'Stonewall' Jackson, his most trusted corps commander, to march west along difficult forest trails and the Orange Plank Road to outflank the Union forces, while he would make a feint at the enemy centre line.[5] On 2 May, Jackson's II Corps marched, while Lee demonstrated as he had planned.

Lee and Jackson's brilliant and audacious flanking manoeuvre at Chancellorsville in 1863 enabled a smaller force to defeat a larger one.

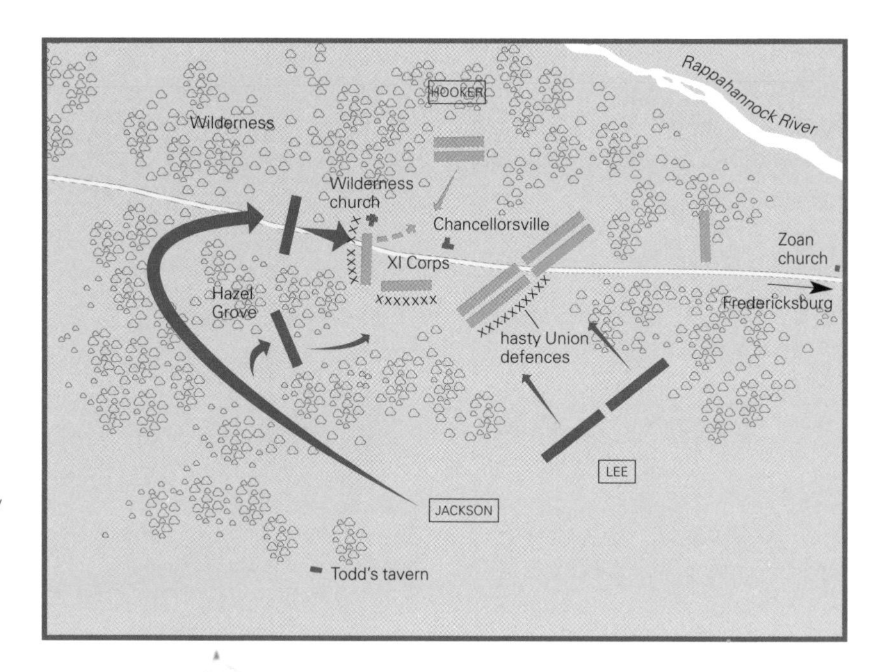

Major General 'Fighting Joe' Hooker rose to command the Army of the Potomac on the basis of his aggressive style, but he was outclassed by Lee.

At 1715 hrs, the Union XI Corps on the right wing was suddenly surprised by the 'rebel yell' of thousands of Confederates pouring out of the trees towards them. The Confederate left swarmed over the log defences and threw the Union forces back. Gunfire from Jackson's batteries was the signal for the renewal of the attack by Lee. Jackson's gun batteries were mounted on high ground at Hazel Grove, which allowed them to rain fire down on to the Union positions in the centre. The area around Chancellorsville house seethed with hand-to-hand fighting and close-quarter musketry, but despite the desperate nature of the defence, the collapse of the Union flank had already determined the outcome. The Union forces were confined to a narrow salient, preventing them from bringing their greater numbers to bear. On 3 May, the Confederates reunited their two wings after further hard fighting. The following day, when Union forces were repulsed at Salem Church 4 miles (6 km) to the east, Hooker realized that his plan had failed and he withdrew. Lee had achieved a remarkable victory.

Analysis

At Bouvines in 1214 Otto IV had attempted a flank attack on the French army which failed because of Philip's bold manoeuvres. Philip's decision to make a series of disruptive charges at the outset of the battle bought him enough time to build up his forces and strengthen his flanks. The sudden collapse of the allied right created the opportunity to attack the vulnerable flank of the allied centre. At Chancellorsville, Lee risked dividing his forces to carry out a flank attack on the more numerous Union forces. The speed and violence of Jackson's attack, when it came, overwhelmed the Union right and, despite all attempts to restore the situation, created the conditions for an overwhelming battlefield victory.

6

Horice, 1423
Leuthen, 1757

Dominating the Terrain and Using the Environment

The astute siting of an army and the intelligent use of the terrain have often been key elements in the outcomes of battles. Concealment using the features of the ground, or the cover of certain weather conditions, can allow an army or a fleet to attack from a new or unexpected direction. The origins of this ruse are lost somewhere in humanity's unrecorded early history – Stone Age hunters knew the importance of wind direction and camouflage when it came to getting close to their prey. However, something of this grasp of deception and the need to make use of the terrain is evident in records in the archives of the Assyrian emperor Sargon II.[1] In his struggle with the Urartians in the 710s BC, he wanted to avoid the obvious approach to his enemies and so he marched northeastwards into the Caucasus. Crossing the eastern end of that range, he swept out of the mountains with complete surprise at Lake Urmia. The concealment of his army's march enabled him to destroy the Urartians decisively.

Battlefield victory is also possible when the terrain is used to enhance an army's defensive stance. In 480 BC the Greeks assembled their fleet in the confined waters between the island of Salamis and the coast of Attica to offset Persian fleet superiority both in numbers and seamanship. And in the 14th century the Scots did not have troops of the quality of England's trained and armoured knights or semi-professional men-at-arms, but they knew their own lands. At Bannockburn (1314), they exploited woods, marshland and a ravine to deny their opponents any room for manoeuvre.[2] On the crowded battlefield, the English knights could not gather the momentum for a charge: struggling through boggy ground and incapable of shock action, they were driven back and lost heavily in a close-quarter battle.

6

Previous pages: Battles in 18th-century Europe were fought *en masse* at close quarters, but the environment could still be crucial to the outcome, as here at Leuthen in 1757.

The deployment of troops on reverse slopes (the rear of a hill) not only prevents an enemy from obtaining any clear idea of dispositions, but also presents fewer targets. This was the preferred technique of the British general, the Duke of Wellington, in his European campaigns between 1809 and 1815, in order to reduce the impact of French gunnery. He was meticulous in his 'reading' of the ground and confessed that he had spent much of his military career endeavouring to find out what was happening 'on the other side of the hill'.[3] Since the advent of advanced battlefield surveillance and satellite observation, the question of terrain may seem less significant, but, in fact, at both the tactical and strategic level, the deployment of troops to take advantage of the ground is still important. Modern gun batteries continue to deploy in re-entrants (valleys or deep folds) because of the cover they can provide against counter-battery fire, and infantry prefer to dig in on reverse slopes in order to 'silhouette' an attacker as they crest the higher ground so their weapons can be used to maximum effect.

Rivers and bridges, hills and passes, and islands and seas have all provided significant barriers and choke points that in consequence become strategically or tactically important: in 480 BC the Greeks relied on the narrow passage at Thermopylae to offset the massive Persian numerical superiority. In c. 1457 BC, in the earliest battle for which we have a record, the Egyptian pharaoh Thutmose III led his army against a coalition headed by the king of Kadesh, whose forces were camped around the strategic city of Megiddo. Of three routes the Egyptian army could take to reach Megiddo, Thutmose chose the most difficult and riskiest – a narrow mountain pass – against the advice of his generals. But it was a successful gamble, as this was also the route his enemies least expected him to take, enabling him to achieve an element of surprise. His ensuing victory at the Battle of Megiddo was recorded in inscriptions on the walls of his temple at Karnak. Equally, environmental conditions can be critical to the outcome of a campaign. Manoeuvres behind banks of mist and fog, as Hannibal exploited against the Romans at Lake Trasimene (217 BC), or movements under cover of darkness to assault drowsy opponents, have at times been decisive. However, to exploit these advantages to the full, troops must be well trained and planning meticulous.

The winter environment has also played a role. In the pre-industrial age, winter campaigns were difficult to sustain, in part because supplies were scarcer, but there could be unexpected benefits. When campaigning against Denmark in the 1620s, Charles X of Sweden was able to cross the sea belts between Jutland and Copenhagen because they had frozen in an exceptionally hard winter. By contrast, Napoleon's Grande Armée was decimated in its 'Retreat from Moscow' (1812) by an over-extended supply chain and Russia's freezing climate (see p.182). In the Second World War, Soviet troops were able to take advantage of the fact that German vehicles and weapons functioned poorly in sub-zero temperatures.

In mountainous country, possession of the higher terrain is vital both to observe and to bring fire to bear against an enemy. Being caught in a defile by an ambushing party has frequently meant disaster. The British-led Indian army on the North West Frontier developed the technique of 'Crowning the Heights'. This involved seizing the highest points and manning them with armed piquets to dominate the valleys and protect the slow-moving columns of troops and baggage below, tactics which in 400 BC Xenophon's 10,000 had deduced for themselves while traversing Armenia. In the Ambeyla Campaign (1863) a British and Indian force was halted by overwhelming numbers of Pathan tribesmen, but their possession of the highest crags, which were reinforced with boulder walls to form 'sangars', enabled them to hold off repeated attacks.[4] These lessons were passed on into the post-colonial era. In the Kashmir conflict between India and Pakistan in the 1980s, attempts to seize the vital ground along the disputed 'Line of Control' led to fighting on the Siachen Glacier, the world's highest-ever battlefield.

Horice, 1423

A classic example of a well-selected position to enhance the advantages of a defending army, which, in turn, gave rise to victory in battle, can be found in the Hussite wars of the 15th century. The followers of Jan Hus, the religious reformer, took up arms in 1415 against the Catholic authorities when their leader was condemned to death as a heretic and burnt at the stake. The rebels seized Prague but were unable to hold on to the city, partly because of their

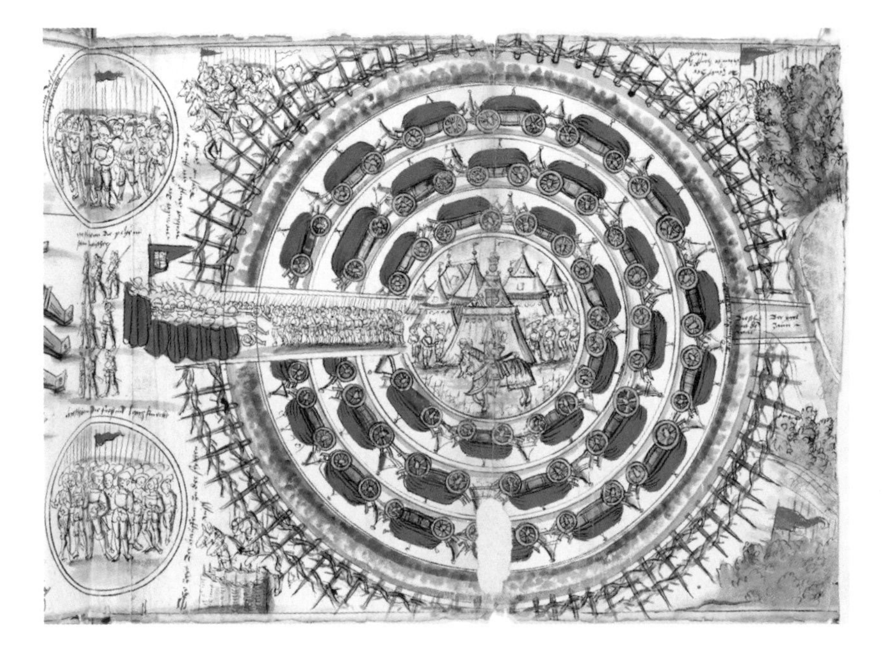

The *Wagenburg* (wagon fort) on the summit of a hill gave Hussite infantry the ability to defeat the Ultraquist army.

own internal splits.[5] Nevertheless, the enigmatic Hussite commander, Jan Žižka, withdrew to southern Bohemia and sustained resistance for some years. By 1422, Žižka had defeated the incursions of the army of the Holy Roman Emperor. With a common enemy removed, the various Hussite factions now began to fight each other. Žižka's zealous Taborites refused to accept the accession of a new Bohemian monarch in 1423. Bohemian Ultraquist nobles and the citizen militia of Prague, together with other Hussite rebels, then confronted Žižka's army at Horice.

Žižka selected the ground for the battle with great care. His main strength lay in his infantry, some of whom were armed with handguns whose slow rate of fire meant that the men needed protection. They were also unable to withstand cavalry charges unless densely packed, but this made them vulnerable to artillery fire. The solution lay in a hilltop position whose steep slopes would slow attacking cavalry to a walk or deny them an approach altogether, and keep Ultraquist guns at a distance, thus reducing their effectiveness.

Žižka deployed his men in a *Wagenburg* (wagon fort). This consisted of a ring of 120 specially designed wagons which provided mountings for cannon and had thick wooden walls to protect the infantry from archery and cavalry. The wagons were secured in place by chains and positioned at angles to enable draught horses to be harnessed to them rapidly. On the move, the wagons could act as personnel carriers, transporting 16 to 20 foot soldiers armed with pikes, crossbows, handguns, flails and shields. Ditches were dug in front of the wagon walls, while gaps between wagons were also useful 'ports' for cannons and handgunners to fire through. Projectile weapons were used to break up infantry assaults or to bring down cavalry horses thus creating more obstacles. The *Wagenburg* was thus a relatively mobile defensive field fortification, but its careful siting was critical.

As expected, the Ultraquist guns found it difficult to bring their fire to bear on the Taborite army on the hilltop, and the cavalry were unable to deliver a charge, so it was clear that they would have to attack on foot. Tired and disrupted by having to climb the steep slopes, the Ultraquists were subjected to close-range fire from the *Wagenburg*. Several attempts to pierce the ring of wagons failed, and, as the assaulting troops began to flag, Žižka recognized the moment had come to unleash his reserves from inside the fortification. A combined force of infantry and cavalry, safeguarded by the *Wagenburg*, swept the Ultraquists back down the slopes and drove them off. The battle was decisive, concluding the Hussite civil war.

Leuthen, 1757

During the Seven Years War (1756–63) Frederick the Great of Prussia faced invasion by Austrian, French, Russian and Swedish armies, but was able to convert this threat into an opportunity. He was committed to offensive action and aimed to march against his enemies before they could combine or converge on his capital, Berlin. Frederick's greatness was not simply this audacity, but his willingness to adapt, to seize opportunities and to prioritize mobility and rapid rates of fire. His emphasis on speed allowed him to redeploy his men against the enemy on one flank, before slower-moving opponents could change front, while his other flank acted as reserve to the main attack. Frederick argued that this 'Oblique Order' manoeuvre enabled him to defeat much larger armies, and, if things went wrong, to preserve a significant proportion of his army.[6]

Sensing that war was imminent in the summer of 1756, Frederick took the initiative and his speed allowed him to secure individual successes against the Saxons and Austrians, but a reverse at Kolin early in 1757 suggested that the weight of the coalition against him was beginning to tell. Yet Frederick attempted to remain on the offensive and, in spite of a raid against Berlin, he defeated a French army at Rossbach in November, concealing his movements behind some hills and then surprising the French – who believed him to be retreating – with a cavalry charge and powerful artillery fire. Flushed with this victory, Frederick arrived in the vicinity of Leuthen in December 1757, where he was confronted by an 80,000-strong Austrian force drawn up along a front of 4 miles (6.5 km) astride his line of march. Frederick's own force numbered no more than 36,000 men and he knew that victory would depend on intelligent use of the terrain and rapid manoeuvre. 'I shall attack, against all the rules of the art,' he wrote, 'I must venture upon this step or all is lost; we must beat the enemy, or all perish before his batteries.'[7]

Frederick's first movement was to feint against the Austrian right. He then took the village of Borne and, when the morning mist had cleared, had a grandstand view of the extended Austrian lines. Better still, his attack had persuaded the Austrians that this would be the direction of the main attack. The rolling countryside aided this misconception, for Frederick was in fact marching the bulk of his force around the Austrian left using a line of low hills to conceal his movements. The Austrians stood and waited, and even when movement to their front was spotted, one Austrian officer thought it was a Prussian retreat: 'The Prussians are off – don't disturb them!' The hidden Prussian regiments were instead marching on to the Austrian flank 'with such precision they seemed to be at a review'.[8]

Frederick drew up his well-drilled infantry almost at right angles to the Austrian line. Although by now they had sensed the danger, the Austrians

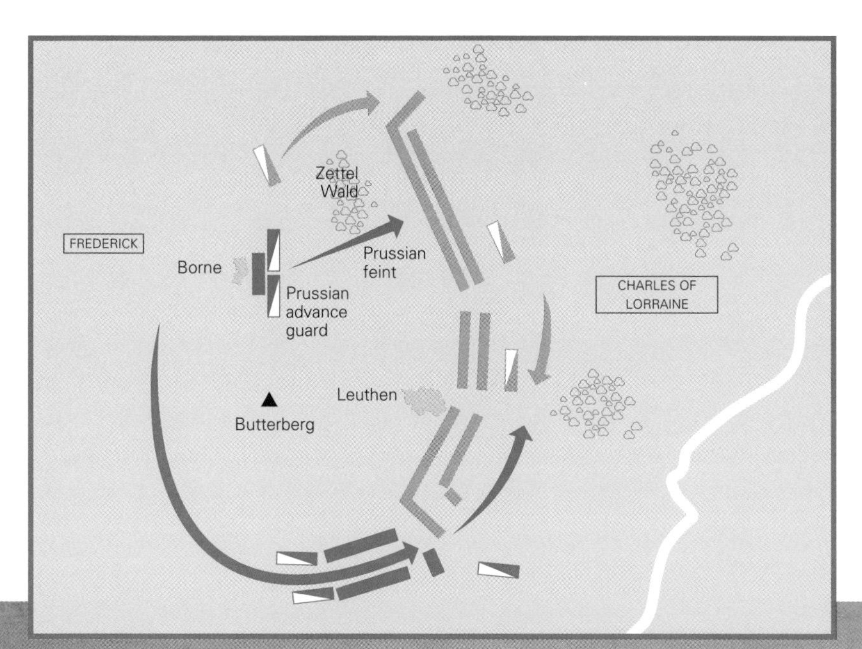

Frederick the Great's use of terrain to cover his advance gave him victory at Leuthen in 1757.

Prussian infantry, Frederick's 'walking batteries', were respected across Europe for their discipline and skill at arms.

struggled to realign themselves against this new threat. Precisely at this moment of confusion, the Prussian infantry assault began. Their firing was so effective they were nick-named 'walking batteries'. Napoleon later wrote that: 'The rapidity in loading is such that it can triple the fire of all other troops.'[9] The Austrian left flank collapsed and fugitives streamed towards the rest of the army. The village of Leuthen became so crammed with troops it was impossible to manoeuvre, and hundreds fell to Prussian round shot.

Nevertheless, Frederick had to call forward his left flank, which had been held in effective reserve, in order to take the position. To push forwards beyond the village, Frederick brought up and massed his heaviest guns. When the Austrian cavalry attempted a counter-attack, they were enveloped on three sides by the Prussian horse. As the Austrian cavalry gave way, the rest of their army started to retreat to the east. Between 7,500 and 10,000 Austrian troops were killed or wounded, and a further 20,000 were taken prisoner. Napoleon described the Prussian success at Leuthen as 'a masterpiece of movements, manoeuvres and resolution'.[10] J. F. C. Fuller noted: 'at Leuthen, Frederick moved, concentrated, surprised and hit. Co-operation was perfect, and so were the dispositions of the three arms.'[11]

Analysis

The landscape of a battlefield has often been used to conceal movement and to protect an army from fire. The selection of the most strategically valuable ground can deprive the enemy of the ability to use their advantages, such as greater numbers or superior weapon systems. At Horice, Žižka used steep terrain to prevent the Ultraquists from applying their main strengths, artillery and mounted knights, forcing them to attack on foot against the walls of his *Wagenburg*. Dead ground has also been useful to conceal movement, and hill features enabled Frederick the Great to redeploy in a defensive stance or to make an attack with great effect respectively at Rossbach and Leuthen in 1757.

Knowing the local terrain has also proved to be a distinct advantage in war. Jungles, mountains and even urban environments provide concealed routes and fire positions. A thorough grasp of the layouts of individual streets and strength of various positions can prove critical. In mountain warfare, possession of the high ground is essential for observation and to dominate the area with fire. With the advent of air reconnaissance, this requirement diminished to some extent, although local protection still requires some presence at higher elevations. Air power has made it more important to be concealed, regardless of the landscape.

Commanders have always had to stay alert to the opportunities afforded by the weather or the terrain. The environment enabled some to create deception plans – Sun Tzu's work is rich with ruses of this nature, such as raising dust to convince an enemy of a particular movement – but it is clear that any operations of this type require well-trained troops who are able to function efficiently in any conditions, day or night, and in any terrain.

7

Leuctra, 371 BC

Echelon Attack

An attack in echelon involves the focus of maximum impact on one particular area of a battlefield, often the centre of gravity, with other sections of a battle-line being withheld or allocated a defensive role. The effect of the echelon attack on a particular area can be strengthened by the application of reserves, which might be deployed as 'waves' of assault troops, either reinforcing or opening fresh fronts along a line or angle of attack, creating a series of hammer blows and denying the defenders a chance to recover. The aim is to maintain a relentless continuity of pressure and concentration of force on one zone of the enemy's line.

In the Iran-Iraq War (1980–88) the Iranians made use of human waves to sustain the momentum of attack at particular points. The Iranians lacked the sophisticated weaponry possessed by the Iraqis but had an advantage in manpower; propelled by a religious and ideological zeal, they hurled themselves at the Iraqi defences. Survivors testified to the way that young Iranians would charge straight into minefields and certain death to clear a route for their comrades behind.[1] In poor weather or in darkness, these seemingly suicidal tactics succeeded in overrunning Iraqi defences on a number of occasions, albeit at a terrible cost. Indeed, on some fronts, Iraqi troops were terrified by the Iranian actions and gave way prematurely.

Previous pages: Echelon attack – a combined Greek hoplite and chariot attack overwhelms an enemy line; Attic vase, *c.* 530 BC.

Leuctra, 371 BC

In the Greek world the standard infantry battle was a simple affair. Contingents of heavily armed citizen-warriors, known as hoplites, lined up opposite each other, often drawn up eight deep, on relatively flat ground and then charged. After the forces collided, a phase of shoving and thrusting with spears and prodding with swords would result in one side losing cohesion and abandoning the battlefield. Tactics were rudimentary: the best warriors tended to be located on the right wing, where their sword arms, while not protected, were also not obstructed by a neighbour's shield, and there was a recognized tendency for armies to drift to the right to outflank the opponent's left wing. Little scope existed for generalship, and commanders usually occupied a place within the line of battle. Although battles might result in limited victories for the respective right wings, there was often no clear outcome unless the victorious wings regrouped to fight a second phase.

Hoplites, the citizen-soldiers of the Greek world, fought in close formation and were armed with a large shield, sword and a spear, which would have been held in the raised right arm of this figurine.

In the 4th century BC, Sparta's domination of the Greek mainland was increasingly threatened by Thebes, its former ally. The Spartiates – males with full citizenship – were renowned as the toughest warriors in Greece, but the allied contingents who comprised the bulk of the Spartan army had little enthusiasm for helping to increase Sparta's power at their own expense. During the 370s, the Thebans prepared for confrontation with Sparta by reorganizing their élite Sacred Band of 300 – whose *esprit de corps* was enhanced by homosexual bonding – improving the physical training of all their citizen soldiers, and building up a self-confidence which had formerly been the prerogative of the Spartans. They were, however, hampered by the reluctance of their local allies in Boeotia to adopt such improvements, since for them Theban successes meant tighter subjugation.

The Thebans had for some time been experimenting with deep formations which might provide an advantage in the initial clash and subsequent shoving in the hoplite engagement: at the Battle of Delium in 424 BC, for instance, the Theban right wing, drawn up 25 deep, had smashed the opposing Athenian formation. Thebes was also fortunate in possessing two leaders of exceptional qualities. Pelopidas was responsible for developing the Sacred Band, which had proved its worth in a small engagement at Tegyra in 375 BC. Here, supported by a few cavalry, members of the Sacred Band had been surprised by a substantially larger Spartan force barring the direct route to Thebes. Instead of retiring to avoid a confrontation, they had charged the Spartans, with the cavalry in the lead, and achieved speedy success by killing all four Spartan leaders. This engagement demonstrated that the fighting quality of the Sacred Band at least equalled that of the Spartans, and thereafter the Thebans deployed its members as a single unit whereas they had previously been spread along the front of the battle-line. Pelopidas' friend, Epaminondas, is renowned as the architect of Theban victory at Leuctra in 371 BC although he was not in overall command of the army, being merely one of the seven annually elected Boeotarchs, the chief officers of the Boeotian Confederacy.

By 371 the Spartan king, Cleombrotus, commanded an army consisting of about 10,000 infantry and 1,000 cavalry, around a quarter of which was Spartan, with the full Spartiates numbering 700. He was under some pressure to fight, since he had built up a reputation for evading opportunities for action. The pressure on Thebes was even greater, as they risked watching Cleombrotus gradually subjugate the Boeotian cities under Theban control, before turning on Thebes itself.

The two armies met on the plain of Leuctra, in the southwest of Boeotia. The Thebans were outnumbered, fielding perhaps no more than about 7,000 hoplites, although in terms of the individual national contingents the Theban units probably matched the Spartan element and were superior in cavalry. Epaminondas' plan, which he had to persuade his fellow Boeotarchs to accept, was to concentrate the well-trained Theban troops against the Spartans. On the assumption that the Spartans would take their traditional place on the right of their formation, this meant locating the Thebans on the left wing where they were drawn up 50 deep. The battle was to be decided by a direct confrontation between these two key contingents.

Four somewhat different accounts of the battle survive: Xenophon, writing at the time, was pro-Spartan and found it difficult to explain the disaster which befell his adopted country; Diodorus, writing some 300 years later, was prone to decorate battle narratives with standard elements; Plutarch in his biography of Pelopidas naturally focused on his honorable actions; and Pausanias in the mid-2nd century AD tried to preserve a Theban perspective. The first phase involved the cavalry, with the Theban horse driving their opponents from the field and possibly disrupting some of the infantry preparations. It is likely that Cleombrotus, as soon as he realized that the Thebans were forming *en masse* immediately opposite him, attempted an outflanking manoeuvre that would have permitted the Spartans to attack the extremely deep Theban formation from the side, and had deployed his cavalry to screen this. Whatever his plans, they were disrupted by the rout of the cavalry and an immediate charge by the Thebans, which was probably angled to hit the Spartans as they deployed further towards the right.

Spartan resistance was robust, even though Cleombrotus was an early casualty. It is likely that they were able to redeploy sufficiently for them to attempt to attack the Thebans in the flank, but this threat was annulled by the Sacred Band charging the Spartans in their flank. Concerted Theban pressure and the deaths of several prominent Spartiates caused the survivors to fall back towards their camp. About 1,000 Spartans died, including 400 Spartiates.

The Boeotian allies do not seem to have contributed to the victory. Although the very rapid advance of the Theban left is sometimes used to explain the failure of the rest of the line to engage, the fact that the fighting around the Spartan units was intense means that there would have been ample time for the Boeotian contingent to close with the Spartan allies if that had been Epaminondas' intention; it is therefore reasonable to assume that he deliberately held back the Boeotian units.

The true heir to Epaminondas' military thinking was Alexander. At both the Issus (333 BC) and Gaugamela (331 BC) he personally spearheaded an angled attack against what he had identified as a key point in the enemy formation, while allocating to the rest of his line a holding or defensive role. In Alexander's case the attacks were delivered by the Macedonian companion cavalry in conjunction with the crack infantry brigade of Hypaspists. They took place on the right wing, from where Alexander aimed to roll up the centre of the enemy line (while the enemy was being pinned down by the Macedonian infantry phalanx). At Gaugamela in particular he used a diagonal advance to draw his Persian opponents away from the battlefield, which they had prepared for a chariot attack. As his move stretched out the enemy line, he exploited the moment a gap opened up between the opposing units to deliver his attack and win the battle.

Analysis

The principle behind an echelon attack is to enable an army to apply pressure at a particular point, or points, in the enemy line, and so it is a special example of concentration and culmination (see **13**). The impact of the echelon may be strengthened through bringing fresh forces into the action. Such tactics have been used throughout history and into the modern era. In Operation Dawn VIII on 9 February 1986 during the Iran-Iraq war, Iranian forces using waves of of dedicated troops breached the Iraqi lines at several key points and captured the Fao Peninsula in southwestern Iraq. It was a significant, if short-lived, victory. However, the Theban victory at Leuctra and Alexander's at Gaugamela illustrate the most successful methods and effects of the attack '*en echelon*'.

8

Strasbourg, 357
Austerlitz, 1805

Committing the Reserve

One of the principles of war is to achieve one's objectives with an economy of effort in order to be able to retain a reserve force that can meet the unexpected, reinforce a threatened part of the front or press home to certainty a successful operation. Historically, the arrival of fresh reserves gave new energy to the fight, added an impetus to operations, created a favourable psychological effect on friendly forces and augmented the numbers already engaged. Conversely, enemy forces could be severely demoralized by the arrival of reserves. When American troops began to arrive *en masse* in Europe during the latter stages of the First World War, it was remarked that they represented a vivifying 'transfusion' for the weakened Allies. To the Germans, their presence in ever-increasing numbers meant that chances of victory were slipping away.

In any campaign, maintaining or reforming a reserve is considered an essential requirement. If reserves are absorbed, the priority is to create new ones as soon as possible. Committing the reserve is often the most important decision the commanding officer can make, particularly in those 'soldiers' battles' where it is difficult to communicate and almost impossible to exercise control once the troops are in action. In each case, the timing and location of this commitment are crucial.

Previous pages: At the Battle of Austerlitz the critical moment arrives for Napoleon and he unleashes his reserve to achieve a brilliant victory.

Equally, one objective of a commander is to draw in the enemy's reserves, thereby reducing their options. A series of probing attacks, if convincing, can persuade an enemy general to deploy his reserves, as Montgomery achieved through his northern thrust at El Alamein (see **1**). Only then, at the *Schwerpunkt* (critical location or time) of the battle, would the main attack be delivered. Two examples of this combination – drawing the enemy reserve, and then committing one's own at the crucial moment – can be found in the Roman era at the Battle of Strasbourg and during the Napoleonic Wars at the Battle of Austerlitz.

Strasbourg, 357

In the Roman Republic the common battle formation comprised three lines: a front line of younger *hastati*; experienced *principes* in the middle; and a reserve of veteran *triarii* who would engage only if the front two lines failed to secure victory. At Zama (202 BC), Scipio Africanus exploited his multiple lines to maintain the effectiveness of his men during extensive hand-to-hand fighting by feeding in the successive lines of fresh troops.

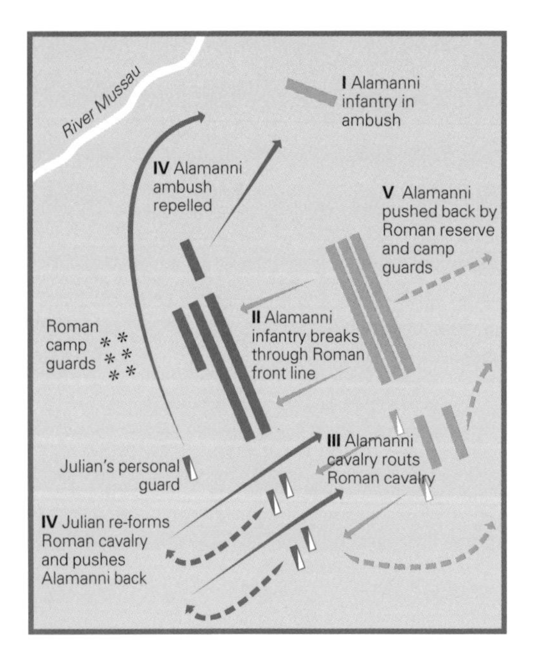

At Strasbourg in 357 initial Alammanic success on the Roman right and pressure on the centre were turned by Julian committing his reserve.

In the mid-4th century AD, the Caesar Julian, whose military experience was very limited, had to confront the Alamanni, a Germanic tribe who had already defeated one Roman army on the upper Rhine near Basel and were marching against the city of Argentoratum (Strasbourg). Julian commanded about 13,000 men and was substantially outnumbered by the Alamanni force, which may have consisted of over 30,000 soldiers, although the Romans had an advantage in cavalry. He deployed his infantry in front of a low hill on which the Roman baggage was placed, in almost antiquarian fashion, with two main lines supported by a smaller third line, flanked on either side by cavalry. The Alamanni placed infantry in an ambush on their right to surprise the Roman cavalry, and located their own cavalry on the left, with the mass of the infantry in the middle drawn up in wedge formations.

A Roman cavalryman overcoming barbarians, on the Arch of Galerius, Thessalonica.

When battle was joined, the death of the Roman cavalry commander on the right and effective attacks by Alamanni light infantry disrupted the Roman units there. On the left, the Romans advanced cautiously and were not caught out by the ambush, which they managed to repel and push back. Meanwhile in the centre, the weight of the Alamannic attack had, after a period of determined resistance, broken through the Roman front line at the point occupied by two of Julian's best units, the *Cornuti* and *Brachiati*. Julian stabilized the situation through a counter-attack from his second line, and the battle became a general mêlée, with the Alamanni having a significant advantage in individual physique as well as numbers, whereas the Romans were better trained and more disciplined. The leading Alamanni with their retinues then mounted a determined charge which forced the Romans right back to the foot of the hill behind their position.

At this point Julian committed the Roman reserve line, bolstered further by the camp guards. After fierce fighting they managed to drive off the Alamanni. The withdrawal turned into flight as the enemy attempted to escape across the Rhine, permitting the Romans to shoot at them as they struggled through the swift current, thereby reinforcing the effect of an already decisive victory.

Austerlitz, 1805

In 1805, the peace that had been brokered between France and the opposing allied coalition partners broke down. Hoping to defeat his enemies before they could concentrate their forces, Napoleon swung his Grande Armée of over 200,000 men across the Rhine and into southern Germany. He surrounded an Austrian army at Ulm and forced it to capitulate.[1] Napoleon

8

then occupied the Austrian capital, Vienna, but not long afterwards the Russians arrived in Moldavia. Having already been compelled to detach troops to protect his southern flank and his lines of communication to France, and to establish a garrison in Vienna, Napoleon was left with 73,000 men to face the combined Austro-Russian army. Having selected a position 70 miles (112 km) north of Vienna to fight a decisive action, his biggest concern was that the Russians and Austrians might refuse battle and draw him into a protracted pursuit. Fortunately for him, the allies were anxious that a withdrawal in winter would be too risky. They also calculated that Napoleon's supply lines would be stretched and an allied victory at this point would force him to conclude a peace settlement on their terms.[2]

A soldier of the French Imperial Guard, the backbone of Napoleon's seasoned army at Austerlitz.

Napoleon chose a deployment that would tempt the allies to attack him on the right. The French were drawn up along a north–south axis, 5 miles (8 km) long, parallel to the River Goldbach. To the south lay marshy ground and a frozen lake, the Satschan. On his northern, left flank he placed a corps under his trusted lieutenant, Marshal Lannes, with cavalry under Marshal Murat. To give the impression of weakness, he withdrew his men from the Pratzen Heights, an undulating plateau in front of the main French position. He pointedly received emissaries from the allied army with excessive courtesy, as if about to seek a negotiated settlement. The key deception, however, was the apparent weakness of his right flank. Here his lines were deliberately thinned. Napoleon reasoned that the allies would attack him there, but, to do so, would need to traverse the Pratzen Heights. Because of the configuration of the ground, the feature concealed the large French reserve, and, at the critical point as the allies became strung out while crossing, Napoleon intended to strike them in the flank.

In all his campaigns, Napoleon's policy was to move rapidly, defeat his enemies' field armies decisively and then impose a political settlement. To obtain flexibility on campaign, he created different *corps d'armée* that operated independently. They could delay an enemy force, combine to deliver the critical blow and disperse to march on separate axes. He also expected his troops to forage wherever possible to reduce reliance on baggage. This flexibility enabled Napoleon to create a reserve rapidly, and to react more quickly in the 'meeting engagement'. His concept of 'central position' was to despatch a corps to pin down and hold off enemy formations that were attempting to combine against him, but to concentrate his corps rapidly to achieve local superiority in numbers. This allowed him to defeat an enemy army before turning to support the corps that had been in the delaying role. Similarly, his *manoeuvre derrière* involved using corps to screen or hold a line, while a larger body swept behind the enemy. At Austerlitz, Napoleon was able to bring up the corps of Marshals Davout and Bernadotte by forced marches, thus significantly augmenting his own force.

The allies drew up their forces, 85,000 strong with 278 guns (almost twice the number of the French artillery), just a mile or two short of the French lines. At 0400 hrs, the allied attack got under way. In the north, the Russians under General Bagration advanced against the French left with the intention of distracting them. Meanwhile, the main force, under General Buxhowden, marched towards the southern end of the Pratzen Heights in a dense mist. By 0700, the French right was under severe pressure, but, just as Napoleon had predicted, the allies were now committed to the attack. As the French checked them, lost ground and then counter-attacked around the villages of Telnitz and Sokolnitz, the allies were forced to push more men forwards into the battle. While Lannes held Bagration, Napoleon ordered Soult forward in the centre. Marching in great columns up the slopes of the Pratzen Heights the French emerged into bright winter sunshine. Before them was the open flank of the allied army, and Soult's troops inflicted devastating casualties on the Russians. The allied army was split in half. As Austrian reinforcements arrived, the French infantry, low on ammunition, gave

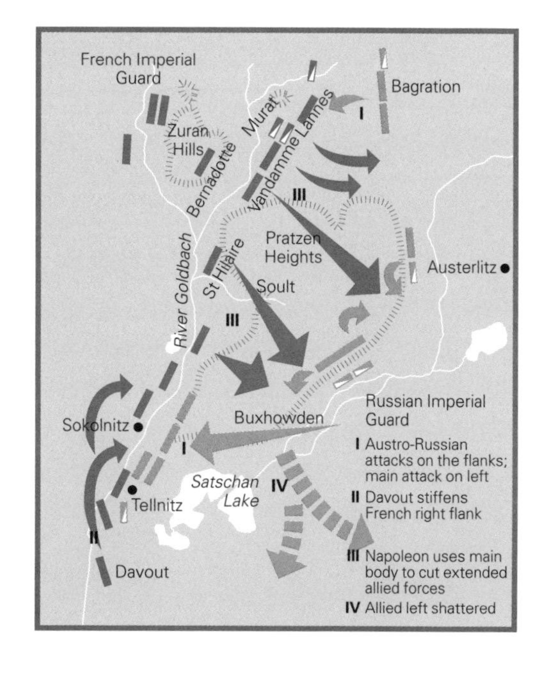

The Battle of Austerlitz in 1805 was decided by the timely arrival of French reserves at the centre of the battlefield.

ground. It seemed they were on the verge of being thrown back down the western slopes, but, despite the odds, they fixed bayonets and charged back on to the plateau. This turned the tide.

To avert a collapse, the Russian Imperial Guard was ordered up the slopes. These men, selected for their loyalty and height, were an imposing sight in white cross belts and towering headdress. Initially they seemed relentless. The French opened fire at close quarters, and eyewitnesses remarked that the Russians fell in whole ranks yet still came forward. Through swirling smoke they pushed Soult rearwards once again. The French Guard Cavalry, their high status denoted by their uniforms of crimson, green and gold, counter-attacked and forced the Russians back. A late Russian mounted counter-charge was also defeated by the timely arrival of Count Bernadotte's 2nd Division. A fresh French cavalry attack broke the remainder of the Russian infantry: the slopes were covered in the dead and dying. Murat's cavalry, operating between the centre and Lannes on the left, now cut the allies off from the Pratzen Heights. Two French infantry divisions, under Vandamme and St Hilaire, continued to exert pressure on the remnants of the Russian force on the eastern side of the plateau.

But an even greater drama was unfolding to the south, for Napoleon was now free to unleash his main reserve – the French Imperial Guard. They marched up and across the Pratzen, turning south towards the spearhead of Buxhowden's force. Unable to move because of the marshy terrain on their flank, the Austrians could not manoeuvre and were compelled to fight on three sides – against Davout to the west and the Imperial Guard to the north and northeast: the result was inevitable. Some Austrian units began to break and attempted to escape across the frozen lake. Hemmed in between the marsh and the French Guard advancing down the slopes, the Austrians had become a disordered mass. French guns on the Pratzen Heights bombarded them with impunity. Some units edged out on the lake. When the ice gave way, order was lost and the Austrian left wing became a hopeless group of fugitives. The French cannonade added to the carnage of the flight. In barely half an hour, Austrian resistance had collapsed.

At 1600 hrs, as darkness fell, the fighting was virtually over. The Russians and the Austrians were back at their starting points, having lost 16,000 casualties and 11,000 prisoners, to the French 7,000. The Austrian emperor, Francis I, called for an armistice in the morning, while the Russians withdrew. Although they were to play a part in further resistance to Napoleon, it was clear the Austrians had been defeated decisively.

Analysis

In the final stages of the Battle of Austerlitz, Napoleon still had several options left, while the Austro-Russian army had none. The difference lay principally in the existence of reserves. Napoleon generally retained a quarter or a third of his force in this role whenever he could. At Austerlitz, he had exercised an economy of effort, using enough men to complete the task of holding the flanks, cutting the allied army in two and then parrying their counter-attacks. Napoleon still had sufficient forces in reserve to cover a withdrawal, or to increase pressure on either flank.

Committing the reserve is always a crucial moment of any battle, designed to avert defeat or to conclude a victory, as was seen at Strasbourg. If the enemy is in the ascendant and the reserve fails to stem the tide, it is likely that events will swing irrevocably against one's own side; indeed this can happen even when the battle is in the balance, since the enemy may well gain heart from withstanding the attack. Many commanders have insisted that, even when reserves were deployed, their first priority was to create new ones. There have also been battles where maintaining or reforming a reserve has been the strategic priority.[3]

At Austerlitz, Napoleon managed not only to retain his own strength, but was also successful at drawing in the enemy reserves. They had been pulled into the fighting on the southern flank and on to the Pratzen Heights; attempts to shore up their wavering formations only served to draw more divisions into the battle. The situation was then difficult to retrieve: Napoleon had already achieved his objective. Austerlitz was won by the precise commitment of men, at the critical moment.

9

Khalkin Gol, 1939

Operation August Storm, Manchurian Campaign, 1945

Blitzkrieg

Blitzkrieg ('lightning war') is a classic manoeuvre of war, involving shock action and rapid penetration into depth to cause confusion and chaos – a combination which can break both the psychological will and the physical capacity of the enemy to fight. The term is particularly associated with the grouping of armour, mechanized infantry and air power used by the German armed forces in 1939–41, although the general idea can be detected in earlier periods. With the introduction of radio communications, however, the speed and distance over which command and control could be extended was increased, while decision-making was devolved and therefore accelerated.

The essential components of *Blitzkrieg* can be summarized as:

- Concentration of force (achieving *local* superiority in numbers or combat power, even if the force is outnumbered in total).
- Extending the enemy's line (to weaken the whole), usually involving deception or some limited deployments that compel the enemy to spread themselves thinly.
- Breakthrough at the point of weakness.
- Race into depth (to engage and pin down or neutralize a mobile reserve, to spread confusion and panic, and to double-envelop enemy positions).
- Maintaining momentum through the breakthrough point, with rapid decision-making and efficient communications.

9

Previous pages: German mechanized forces swept across Europe in 1939 and 1940 in high tempo operations.

There was no specific doctrine of *Blitzkrieg* in the German army at the start of the Second World War; what existed were more nuanced ideas surrounding specific instructions on communications and leadership. The origins of *Blitzkrieg* in German military thinking can be traced to the use of storm troops in the First World War (see p.30).[1] Nevertheless, these were, essentially, infantry tactics and did not change Germany's strategic situation during the latter stages of that war. In 1929, when new ideas were being sought to avoid the humiliation of 1918, armoured warfare doctrine was an attractive alternative. Experiments in combined armoured-air operations were carried out in the western Soviet Union under a special agreement. Put together with the earlier thinking about infantry, which could be mechanized to keep up with the armour, the whole tempo of operations could be increased. However, this concept of *Blitzkrieg* might have remained simply an aspiration had it not been for a revolution taking place within the German army itself.

Hitler, always impatient for success, favoured ideas that would guarantee rapid results. More junior or progressive commanders were willing to use the environment Hitler created to abandon the defensive strategic thinking of the inter-war years. They were prepared to embrace the ethos of *Blitzkrieg* as the basis for operational decision-making and to sacrifice strategic thinking. Thus, when Hitler's armies invaded Poland, the Low Countries and then France, *Blitzkrieg* represented, according to historian Michael Geyer, the triumph of short-term planning and operational management over strategy.[2] Aware that Germany lacked the capacity for a long war, or even for another war of the intensity of 1914–18, Hitler gambled on delivering shock defeats, to persuade his enemies to capitulate before they geared up for a longer struggle. Yet the United Kingdom, United States and Soviet Union possessed precisely the capacity to be able to withstand the shock of *Blitzkrieg*. Moreover, the German army of 1939–41 could only mount such operations with a relatively small portion of its forces: many German troops *marched* into Russia because they lacked the requisite mechanization, and most German guns were still horse-drawn.[3] This contrasts sharply with the Soviets, who developed their own version of *Blitzkrieg*, and the Americans, whose European forces and logistics were entirely mechanized.

Although *Blitzkrieg* was credited as the system that broke the Allied armies in Western Europe in May 1940 near the beginning of the Second World War, the French army had already been weakened to some extent before the German attack began. Many of its members were eager to avoid the sort of bloodbath seen in the First World War and the French high command placed undue emphasis on the Maginot Line defence system. Armoured warfare doctrine was barely developed and the whole French strategic emphasis lay on defence, reacting to enemy initiatives and protecting the entire frontier. During the winter of 1939–40, inactivity further demoralized the French troops. When the attack finally came, the Germans pushed through the relatively poorly defended Ardennes, the weak spot in the French defences. The Germans then rapidly advanced in depth causing utter confusion, and rolled up the Maginot defences from the flank and rear. Turning upwards towards the Channel coast they threatened to cut off the British Expeditionary Force, which had to conduct a fighting retreat to Dunkirk.[4]

Examples of highly mobile forces penetrating into depth to attack enemy logistics can be found in other periods. In the American Civil War, Major General J. E. B. Stuart's Confederate cavalry corps, which had accompanied General Robert E. Lee's army, made circuitous and extended raids through Union territory, both in 1862 during the Jamestown Peninsula Campaign and in 1863 in other parts of the eastern theatre.[5] Stuart gained valuable intelligence and made opportunistic attacks to disrupt the Union army's logistics. However, it was the Union commander Major General William Tecumseh Sherman who exemplified the effects of striking in depth against the enemy's supplies and logistics base. In his famous March to the Sea in 1864, his army cut a swathe through Georgia and South Carolina, consuming and destroying everything in his path. The effects were both psychological and physical. A similar linkage with raiding philosophy can be observed in the example of the Cossacks in the Russo-Japanese War (1904–05). The Independent Trans-Baikal Brigade under General P. I. Mishchenko, comprising 7,500 cavalry, mounted infantry and horse artillery, twice penetrated from Manchuria into the Japanese-held Korean peninsula in 1904.[6] The following year it was tasked with another deep raid, to cut the Port

Arthur–Harbin railway. In the same area the Russians achieved far more significant *Blitzkrieg* effects some 40 years later, as we now discuss.

Khalkin Gol, 1939

The Japanese occupation of Manchuria and Korea, and the invasion of China in the 1930s, gave Imperial Japan and the Soviet Union contiguous borders for the first time. The precise frontier was disputed and the first 'border incident' occurred in 1938 in the Primorye region in Manchuria in what became known as the Battle of Lake Khasan. Some months later, after further minor clashes, a far more significant battle took place on Khalkin Gol River, where small skirmishes escalated into a serious confrontation. On 11 May 1939, mounted Mongolian troops entered disputed territory and were attacked by Japanese Manchurian cavalry. The Mongolians returned in force but were unable to eject the Japanese, who, in their turn, also brought up more men. However, the second Japanese attack, still with relatively small numbers, was enveloped and virtually destroyed. This incident provoked a Japanese air raid on the local Russian airbase. Eager to avenge the slight and to assert the border as they saw it, the Russians built up troops in the area, as did the Japanese. The Japanese developed a plan to envelop the Soviets in a pincer movement, with two wings of brigade strength and made up of both infantry and light tank units. The attack took place on 1–4 July.

The Soviet *Blitzkrieg* counter-offensive at Khalkin Gol defeated the Japanese decisively.

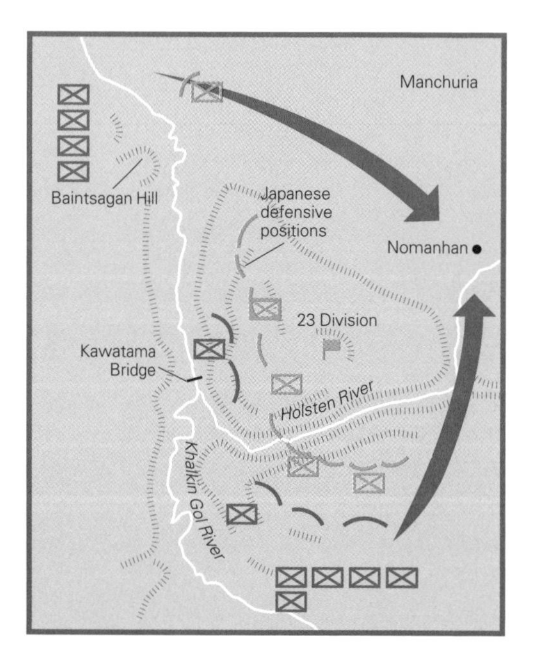

The Japanese attack began well, but they found it impossible to maintain momentum. The northern wing succeeded in crossing the Khalkin Gol and seized Baintsagan Hill. However, Lieutenant General Zhukov, who was later to enjoy a famous career in the Second World War, intercepted this wing and counter-attacked with over 400 armoured vehicles. Blunted by this parry, the Japanese spearhead withdrew back across the Khalkin Gol. The southern Japanese wing attacked the Soviets at night, but, despite a bitter struggle, the Japanese were unable to break the Soviet formation or take the Kawatama Bridge. They too withdrew. Operating on very extended supply lines, both sides took two weeks to build up sufficient men, munitions and equipment to make another major attack. On 23 July, the Japanese began an assault with a heavy artillery bombardment, but made limited progress, suffering 5,000 casualties; with dwindling ammunition, they were forced to withdraw a second time.

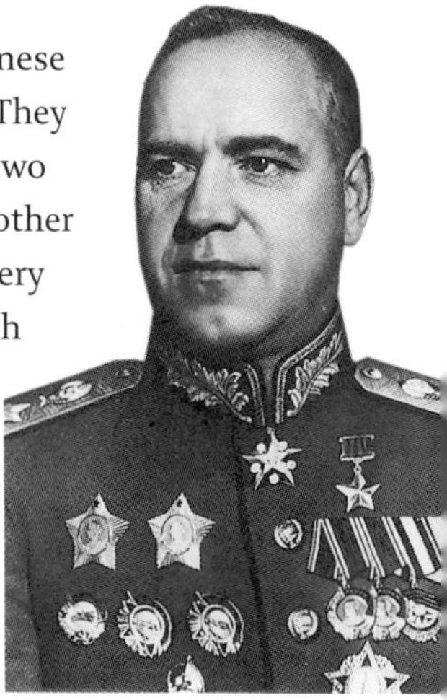

Zhukov had amalgamated three tank brigades and two mechanized brigades into an armoured strike force to augment the three rifle divisions, two tank divisions and two smaller tank brigades he had in the field. Supported by fighter and bomber aircraft, he possessed a strong force with which to take on the 75,000 Japanese and their own air forces in the region. In fact, at the point where Zhukov chose to attack, the Japanese had only two light tank divisions.

General (later Marshal) Zhukov, one of the most talented Soviet commanders, achieved considerable success in Manchuria as well as in eastern Europe.

On 20 August, Zhukov's main force, the bulk of his 50,000 men, attacked and pinned the strongest Japanese formations and the armoured striking force in two wings. The Russians drove into depth and enveloped the Japanese 23rd Division – their speed enabled them to penetrate deep into the Japanese rear. The Japanese made several attempts to cut their way out of the encirclement, and to relieve the 23rd Division, but they failed to make headway against Russian artillery and air attacks, or the substantial Soviet armoured forces. Japanese losses were heavy, perhaps as high as 45,000, although the official statistics are much lower. The effects of the battle were certainly far-reaching: the Japanese decided to abandon the idea of striking into Russia to acquire resources and focused on the second option of attacking Southeast Asia and neutralizing American power in the Pacific. A negotiated settlement meant that the Soviet Union did not have to fight on two fronts in 1941.

Operation August Storm, Manchurian Campaign, 1945

The scale of the Soviet offensive in Manchuria in August 1945 was overwhelming. One and a half million men were assembled in three Red

9

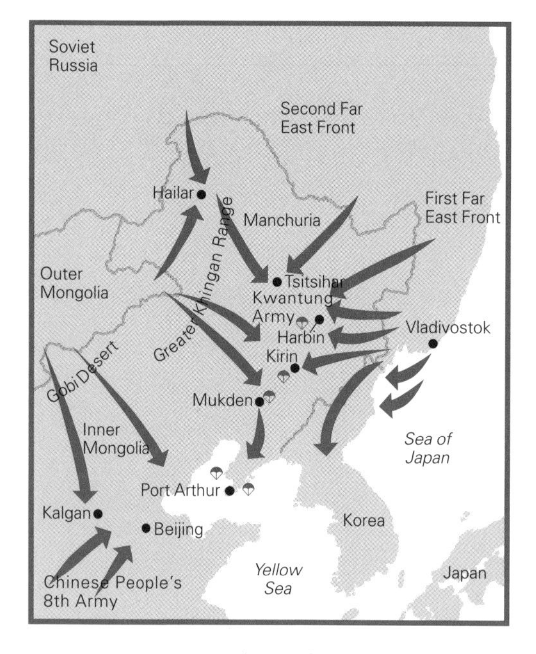

August Storm, the climax of the Soviet Second World War land campaign in Asia in 1945.

Army Fronts, bringing together 12 armies and 2 air armies. There were over 3,700 tanks, 28,000 guns and 4,300 combat aircraft. The Japanese Kwantung Army's forces in Manchukuo (the name of the Manchurian puppet state) numbered no more than 600,000 with 1,215 armoured vehicles, 6,700 guns and 1,800 aircraft (many of which were obsolete or unserviceable). Much of the Kwantung Army had spent the period after 1941 in a counter-insurgency role, and the bulk of the force was made up of new recruits, since more experienced men had been drawn into the Pacific theatre or the occupation of Southeast Asia. Japanese military intelligence had failed to forecast the Soviet moves in the summer of 1945, including the enormous logistical effort of transferring such a large force across the deserts and mountains of Mongolia, so most of the Japanese forces were deployed in the south.

The Soviet attack began on 8 August 1945, the main thrust being across the Greater Khingan Range into central Manchuria. Japanese forward units found it impossible to communicate because air and artillery attacks on them quickly neutralized command and control centres. Large Soviet armoured formations, using tactics developed during the fighting in the west against the Germans, swept around Japanese positions and drove into depth, enveloping the main Japanese formations in a giant pincer. Airborne units were dropped ahead of the main thrusts to secure city centres and airfields. Air forces also delivered fuel to the forward armoured units to maintain momentum. Within a week, the Soviets had penetrated deep into Manchuria, captured the principal cities and seized the Emperor of Manchukuo.

The terrain lent itself to the *Blitzkrieg* philosophy – the large expanses of open country were ideal for tanks and air observation – but the Soviet operations in Manchuria were a success because of the scale of their forces and their proficiency in these combined operations.

Analysis

Whether *Blitzkrieg* was a clearly developed doctrine in 1939 or not is, in a sense, immaterial when one analyzes the success of the Wehrmacht or the Soviet Army during the Second World War. The principles or ethos of the system were embedded within the thinking of the personnel involved. The essence of *Blitzkrieg* was the shock effect it produced in the minds of the enemy. Success relied on maintaining momentum, the continual exploitation of opportunities and the speed of operations. Where all ranks knew they must avoid delay, drive into depth along coherent axes and support these thrusts with sufficient firepower and air support, then success was much more likely. When the spearhead was held up, intimate fire support from tanks or aircraft could sometimes prove sufficient to restart the advance, but good communications and the ability to shift the axis of advance rapidly so as to swing around any strong point were equally important.

Blitzkrieg also depended on deception. The enemy was deceived as to the main point of attack and often lulled into a false sense of security (either of their own making or encouraged by the attacking force). Much preparatory work was required in terms of intelligence gathering (on enemy troop strengths and dispositions) and logistics. When the attack was delivered, neutralizing enemy logistics bases and blinding their command and control nodes rendered them ineffectual and immobile. The attacking forces were able to anticipate their enemy's 'decision-making cycle' by devolving their own decision-making and ensuring that all ranks understood the importance of initiative in fulfilling mission tasks, and not waiting for orders. This concept the Germans labelled *Auftragstaktik*, and it gave mobile armoured forces the ability to effect changes before the enemy could react.[7] By the time an opponent issued fresh commands, the situation had changed. These factors were as important as, if not more so than, the technologies involved.

10

Carrhae, 53 BC
Omdurman, 1898

Concentration of Firepower

Ever since the invention of projectile weapons, it has been possible to overwhelm an enemy with superiority of fire rather than numbers or close-quarter fighting. Through the ages, developments in the greater range, accuracy and lethal impact of projectiles have, ultimately, led to the intercontinental ballistic nuclear missile. A single missile is capable of carrying a devastatingly destructive warhead or, indeed, multiple warheads, each of which could wipe out a city. At various points in history, armies have sought to bring to bear the maximum firepower they can, to inflict as many casualties as possible and to 'soften up', neutralize or destroy entirely their adversaries.

In the ancient world, the javelin or bow served this purpose. Well-trained forces could launch their projectiles with great accuracy and achieve high rates of fire. Javelin-throwing peltasts under the Athenian general Iphicrates wiped out a whole Spartan brigade at Lechaeum in 392 BC. Some forces perfected the art of archery from the saddle. The famous Parthian shot was delivered by horsemen who, while riding away from the enemy, twisted round in the saddle to shoot before their opponents could engage them. Armies sometimes employed archers *en masse*, hoping to create a barrage of arrows either on a relatively horizontal trajectory or in a shower, plunging fire on to the heads of their opponents. The Mongols favoured the smaller Central Asian composite bow used from horseback, while the English and Welsh longbowmen, whose archery devastated French armoured knights at Agincourt in 1415, relied on a large yew bow. Its 30-in (50-cm) arrows could be shot by experienced archers on foot with a force of 150–200 lb (68–91 kg): their penetrative power explains the archers' fearsome reputation in medieval Europe.[1]

10

Previous pages: The overwhelming power of the breech-loading rifle and machine gun: British infantry at Omdurman in 1898.

The invention of gunpowder weaponry led to new possibilities on the battlefield. Arquebuses, muskets and the first rifles were rather short-range weapons and often unreliable, particularly in poor weather. However, used *en masse* they could create a significant effect, which increased as weapons became more sophisticated, making it impossible to remain exposed or grouped together within firing range without suffering severe losses. Firepower not only favours the defence – a suppressing barrage laid down ahead of advancing forces enables them to move forwards and close with an enemy. Infantry developed systems, or drills, for the operation of weapon systems and also to co-ordinate movements and the fire of mutually supporting fire teams. Despite anxieties in the mid-19th century that higher rates of fire would exhaust ammunition supply in any prolonged battle, the sheer intensity of fire could suppress or destroy enemy units far more effectively and quickly than previously, a trend which continued with the advent of the machine gun.[2]

In fact, commanders throughout history often attempted to create a 'killing area' using their projectile weapon systems. The aim was to establish a zone into which an enemy could be lured in order to maximize the effect of fire. At Carrhae, as at Omdurman almost 2,000 years later, the combination of a 'killing area' and the sheer weight of fire were key factors in producing victory.

Carrhae, 53 BC

In 53 BC, in Rome's war against Parthia, Marcus Crassus, the Roman governor of Syria, crossed the Euphrates at Zeugma with a substantial army of seven legions. His force numbered 35,000 heavy infantry, plus 4,000 cavalry and 4,000 lighter infantry; his intention was apparently to march on the Parthian capital of Seleucia. He was also keen to co-ordinate actions with allies in Armenia and to confront whatever Parthian forces were in the vicinity.[3]

Reports of the proximity of Parthian troops enticed Crassus forwards, especially when it appeared that they were not numerous, but the Parthian general Surenas had concealed the majority of his men behind an advance guard, and ensured that their bright armour was cloaked. When the Romans were suitably close, the Parthians revealed themselves with a disconcerting racket from tambourine-like instruments and prepared to charge down the Romans with their heavily armoured lancers. The Roman disposition was sufficiently cohesive to deter this, and instead the Parthians resorted to archery from all sides. Crassus arranged his men in a hollow square, with cavalry posted on each side in an attempt to drive the archers beyond effective range. However, any sallies from the Roman square were met with heavy fire, and even if the Parthians were forced to withdraw briefly they continued to inflict casualties on their pursuers using the famous 'Parthian shot'.

Roman soldiers depicted on the altar of Domitius Ahenobarbus of the 1st century BC. At Carrhae their armour and equipment could not save them in the face of sustained attacks from Parthian archers.

In an attempt to force a way to safety, Publius Crassus, Marcus' son, attacked the enemy with most of the Roman cavalry, reinforced with 500 archers and about 4,000 infantry. The Parthians gave way, luring Publius further from the main army until they could safely effect a second encirclement; the detached force was soon under heavy pressure, and in desperation Publius and many of the cavalry perished in a charge against the armoured Parthian lancers.

Although suffering constant losses, the main Roman army continued to resist, partly in the hope that the Parthian archers would eventually run out of arrows. Surenas, however, had provided against this by organizing a train of camels loaded with fresh supplies so that his horsemen could rapidly reload their quivers and return to the fray, maintaining their barrage. In spite of the dismay caused by Publius' failure and the parade of his head on a pole, the Romans managed to hold out till nightfall, at which point the Parthians, following their normal practice, withdrew. Crassus retreated towards Carrhae (modern Harran, in Turkey), but Surenas mopped up Roman stragglers and the surviving troops were again surrounded, and subjected to more intense archery. Crassus was eventually lured into negotiations with Surenas and killed, while the Parthians gained the precious legionary standards.

Omdurman, 1898

The Battle of Omdurman in the Sudan represents the epitome of the clash between a modern industrialized army and a force whose technology and fighting methods had remained unchanged for centuries. The British-Egyptian and Sudanese army under the command of General Herbert Kitchener had been given the task of the reconquest of the Sudan in 1896, following years of the Mahdist rebellion. The Mahdists, or Ansar, had been inspired by the religious rhetoric and charisma of Muhammad Ahmad, the so-called Mahdi, and the rebellion had been sustained by his successor, Abdullah al-Taashi, the Khalifa.

It took Kitchener two years to construct an effective logistics system and a railway into the heart of the Sudan, so that, by September 1898, he could move his troops and was ready to seize the Mahdist capital at Omdurman, near Khartoum. The Khalifa had assembled a force of 50,000 Sudanese

An artist's impression of Omdurman: in reality, the Mahdists were cut down before they could reach the British-Egyptian army.

tribesmen armed with bladed weapons, spears and firearms. They were highly motivated and had previously achieved a succession of victories using their great weight of numbers. The Sudanese desert had been a refuge for them, but had also provided a hostile territory into which less well-equipped or weaker forces could be lured. Unable to stem Kitchener's more methodical invasion, the Khalifa hoped to spring a vast ambush at Omdurman. His forces were divided into five groups. The smallest, numbering 8,000, was placed directly opposite the British-Egyptian army on a low ridge but in full view. The rest of the Ansar was hidden behind lines of hills to the northwest, west and south. The plan was to tempt the British into the space between the hills and then to launch attacks that would come to close quarters, where the Khalifa's numbers would be able to tell against the Europeans.[4]

Winston Churchill, then a lieutenant attached to the 21st Lancers, was part of a mounted patrol observing the Khalifa's army. 'The whole side of the hill seemed to move,' he related, 'and the sun, glinting on many hostile spear-points, spread a sparkling cloud.'[5] This great host, trotting amidst clouds of dust and inspired by hundreds of billowing flags, was concentrated on a frontage of just 4 miles (6.5 km). But Churchill was also aware that, however impressive a spectacle, this army was heading directly for one of the most heavily armed formations of the modern world.

The British-Egyptian army, 25,000 strong and with 44 guns (breech-loading, rifled artillery) and 12 machine guns, was drawn up along the Nile in a shallow arc behind a *zeriba* (enclosure) hedge of thorn bushes around the village of Egeiga. Its flanks were protected by cavalry, while river-boat guns provided additional support from the Nile. Kitchener would have been compelled at some point to move from this position and advance on Omdurman, but the Khalifa's forces were poorly co-ordinated and launched themselves prematurely at the British lines.[6] The first wave consisted of about 16,000 men, some mounted but the majority advancing on foot. The British artillery opened fire at 2,700 yd (2,470 m), inflicting severe losses and breaking up the cohesion of the attack. The tribesmen struggled on, only to be met by the withering fire of rifles and Maxim machine guns.

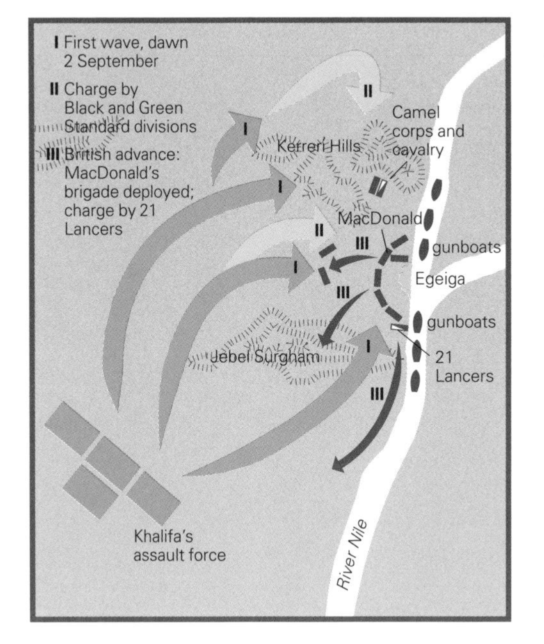

The Battle of Omdurman, September 1898, demonstrated the importance of firepower against overwhelming numbers.

The Khalifa's casualties in this phase of the battle were estimated at 4,000, and the Sudanese had failed even to reach the British-Egyptian lines. A small party occupied a shallow depression about 300 yd (275 m) from the *zeriba*, but despite putting down some harassing fire they were soon neutralized by the British artillery. A subsequent thrust was made by the Ansar from the southwest but this was also cut up, and an attack on the northern portion of the British-Egyptian army barely developed before it was destroyed.

Kitchener now moved off towards Omdurman itself. The British cavalry were ordered to clear the way and the 21st Lancers deployed to charge down what appeared to be a weak skirmish line. In fact, this line screened a force 2,000 strong hidden in a depression. The Lancers lost a quarter of their 400 men in the action, but rode right through the Sudanese and regrouped, and the incident did not halt Kitchener's drive towards Omdurman.[7] The Khalifa had not yet committed the bulk of his army, however, and an opportunity now arose to attack the rear of Kitchener's column. The brigade commanded by Hector MacDonald, comprising mainly Sudanese troops of the Egyptian army, reinforced with Maxims and artillery, was some distance behind the main body.

The Khalifa launched around 15,000 tribesmen of the Black and Green Standard divisions against the brigade, but MacDonald had been alerted to the threat and had sufficient time to wheel his men into a firing line. Although MacDonald's Sudanese troops were equipped with inferior weapons, they were still able to inflict grievous losses.[8] The Maxims and artillery once again caused carnage, and the arrival of support from the main body added to the weight of fire being poured into the Khalifa's flagging assault. The Khalifa's mounted men made a desperate charge, only to be cut down before they could come to close quarters. The Khalifa's northern wing arrived too late to change the direction of events, and they too, in their turn, were mauled by the sheer firepower of the British and Egyptian brigades. The battle had begun at 0600 hrs. By 1130 it was all over and the Ansar had lost 23,000 killed and wounded, with 5,000 captured. The British, Egyptians and Sudanese had lost just 48 men. Kitchener, with considerable understatement, remarked that the enemy had had a 'good dusting'.[9]

Analysis

The firepower of the smaller British and Egyptian forces, like the Parthians at Carrhae, was overwhelming against the dense formations of their adversaries. At Omdurman the open nature of the landscape gave the British excellent fields of fire and they were able to use their weapons at maximum ranges. The Maxim machine gun fired up to 600 rounds a minute, while riflemen, equipped with magazine-fed .303 rifles, could deliver some 15 aimed shots a minute. Smokeless powder propellant, developed in the 1880s, avoided the clouds of smoke that had traditionally obscured the targets, so it was possible, by 1898, to maintain a constant stream of fire. Breech-loading artillery could also achieve high rates of fire and deliver exploding shells whose fragmenting cases sent ragged shards in all directions. The tribesmen lacked technological equivalents to these systems: they were simply outgunned. The historian Bruce Vandervort noted that the Khalifa's plan was for two enormous formations to 'work together to spring a gigantic ambush on the British army', but 'It went awry ... not because it was badly conceived but because of the Khalifa's fatal miscalculation of the enemy's firepower and his insufficiently detailed knowledge of the enemy.'[10] Surenas and Kitchener both understood that victory for their forces could be achieved by putting maximum sustained fire on the enemy.

11

Arsuf, 1191
Balaclava, 1854

Shock Action

At the critical moment of battle, the shock action of a charge, or a brief increase in the intensity of fire, is often enough to break an enemy force. Throughout history, the blow of a sudden assault has frequently, if not exclusively, been delivered by 'heavy' troops, such as infantry or cavalry/armour, as these were designed specifically to punch their way through an enemy line. The impact of a charge – indeed, sometimes the very spectacle of it – can prove too much for the troops on the receiving end, even those trained and disciplined to expect it.

Usually, shock action will only succeed if the enemy line has been 'softened up' or its fire suppressed sufficiently to allow an attack to close with adversaries; otherwise the attack will lose cohesion through casualties in that final short distance to the enemy force. Deception, or engaging the enemy elsewhere, can also enable the attackers to choose the point and moment for the shock action assault. One good example of this is the Battle of Arsuf in 1191.

11

Previous pages: The Heavy Brigade victorious in a desperate charge at Balaclava in 1854.

Arsuf, 1191

In the medieval period mounted knights riding heavy horses up to 15 hands high provided the shock effect. They were protected by an iron cap, usually worn over the hood of a chain-mail shirt or hauberk that extended down to the knees. A kite-shaped shield covered the left side from shoulder to knee. The main offensive weapon was an 8-ft (2.4-m) iron-tipped lance, but for close work knights also carried a sword, axe or mace. This equipment weighed about 66.2 lb (30 kg), to which must be added the stirrups, straps and wooden saddle that clamped the knight to the back of his horse. An animal like this at full gallop would have been an awe-inspiring sight, and any substantial group must have seemed terrifying. As each soldier lowered his lance the full momentum of horse and man would be concentrated in the point. A Byzantine princess commented: 'A mounted Kelt [Frank] is irresistible: he would bore his way through the walls of Babylon; but when he dismounts he becomes anyone's plaything.'[1]

A mass charge of a substantial body of knights could be highly effective, but there were major obstacles to delivering one. It was only possible on an open battlefield, so the site had to be picked carefully. The charge was a once-only technique because the heavily laden horses could make just one short sharp attack. If it was not timed and directed correctly to deliver a critical blow, the knights on their blown horses would be easily picked off. At the siege of Damascus in 1148 a Muslim contemporary described the Franks biding their time: 'The infidel cavalry waited to make the charges, for which it is famous, until a favourable opportunity presented itself.'[2]

An even greater problem was that knights were men of substantial rank who resented discipline, not members of standing armies. In Europe they were accustomed to fighting in small groups, often called *conroi*, because much warfare was local; even in major conflicts armies avoided battle, preferring to raid and destroy, provoking frequent small-group conflict. As a result, the ethos of knightly warfare was highly individual. When a great army gathered, the *conrois* formed small units within the retinues of lords who were in direct contact with the commander, usually a king or nobleman. But their rank was such that they could not simply be given orders. In the absence of any real structure of command, therefore, the overall military leader had to impose himself by sheer force of personality and the example of his bravery.

It was difficult enough for a king to control armies made up of his own people, but crusades were gatherings of many peoples. And the nature of the Islamic armies they were fighting against, and their tactics, presented special problems. The Turks, the dominant military élite in the Middle East, were a steppe people. They were primarily horse-archers whose main tactic was to surround the enemy and bombard him with arrows, then to charge home only when gaps appeared or small units became isolated. More particularly, the Turks tried to shoot the horses of the knights to prevent them making an effective charge, forcing the Franks to send forward a screen of infantry to hold off the horse-archers. But this was no easy task, and the light horsemen presented few solid targets for a Frankish charge. An eyewitness of the Third Crusade commented:

> If they are hotly pursued a long way they flee on very fast horses. There are none nimbler in the world, with the swiftest gallop – like a flight of swallows. It is the Turks' habit, when they realize that their pursuit has stopped following them, to stop running away themselves – like an infuriating fire which flies away if you drive it off and returns when you stop.[3]

The Third Crusade (1189–92) was an attempt by western Europeans to reconquer Jerusalem, which had been lost after the Battle of Hattin in 1187 (see p.178). Its early focus was the city of Acre, which was besieged by a multinational force under bickering leaders from 1189 until its surrender on 12 July 1191. By that time it was obvious that the greatest soldier in the camp was Richard I of England, the 'Lionheart', and leadership of the large but motley host fell upon him. The army now needed to move south to establish a good seaport base at Jaffa, from where it could strike inland to Jerusalem. Many crusaders seem to have felt they had already 'done their bit', so Richard had to tempt them by flattery, prayer and even force; but above all he set an example by moving out into the plain exposed to enemy attack: 'It was the royal custom to be always first to go out armed against the Turks when they attacked, to punish them as far as divine providence allowed.'[4]

By such means Richard established his mastery over the army – an essential preliminary for what lay ahead. Because the 20,000-strong army of his opponent, Saladin, was clearly prepared to dispute his passage to Jaffa, Richard needed to organize his force for a 'fighting march'. This was a stratagem, unknown in Europe, of the Latin East, where it was sometimes necessary to form an army into a tightly disciplined 'moving fortress' to traverse enemy territory. Richard formed his shock cavalry into three squadrons, protecting their exposed left flank with a screen of footmen and archers. The baggage train moved on the right side of the cavalry close to the sea and his accompanying fleet, and infantry formations were rotated among them to give them rest. From the very first day, 22 August, the Muslims harassed the army, attempting to kill horses and break down the infantry screen. One of Saladin's senior staff remarked:

> The enemy army was already in formation with the infantry surrounding it like a wall, wearing solid iron corselets and full-length well-made chain-mail, so that arrows were falling on them with no effect ... I saw various individuals among the Franks with ten arrows fixed in their backs, pressing on in this fashion quite unconcerned.[5]

A 13th-century bronze *aquamanile* (water container) in the form of a mounted knight. European knights dominated the battlefield because of their speed, force and shock effect.

On several occasions the Turks pressed the crusaders, notably on 30 August, when Hugh of Burgundy's force was almost cut off. However, the coastal plain was narrow and Saladin did not dare to commit too many troops to the attack lest they be trapped. But at Arsuf, where the plain widened, his army deployed for battle. Richard reorganized his cavalry into five units, with the Templars in the vanguard and the Hospitallers forming the rearguard. His main objective was to press on to Jaffa, but this redeployment gave him flexibility.

Saladin's attacks focused on the rearguard where the crossbowmen and archers were driven to fighting while walking backwards, and the Hospitallers received attack after attack. Richard refused to permit any counter-attack, concerned to husband his shock-force and perhaps also hoping to ride out the storm and maintain progress to Jaffa.

Nevertheless, under severe pressure, the Hospitallers charged. To support them Richard was forced to commit all his cavalry, which did tremendous damage to Saladin's army. Baha al-Din saw the charge forming up and felt its terrible effects:

> They took their lances and gave a shout as one man. The infantry opened gaps for them and they charged in unison. One group charged our right wing, another our left, and the third our centre. It happened that I was in the centre which took to wholesale flight. My intention was to join the left wing, since it was nearer to me. I reached it after it had been broken utterly, so I thought to join the right wing, but then I saw that it had fled more calamitously than all the rest.[6]

Richard was at pains to rein in his cavalry and prevent them from dispersing, while Saladin and his leaders, despite losses, kept control of the core of their army. As a a result, despite a terrible defeat in which 7,000 died, including 32 emirs, Saladin's army continued to exist. The next day he returned to harass the crusaders, but, as Richard himself said in a letter home: 'Since his defeat that day, Saladin has not dared to do battle with the Christians.'[7] Arsuf shows the effectiveness of shock action in the form of the massed cavalry charge.

Balaclava, 1854

During the Crimean War (1854–56), a joint British-French-Turkish expedition besieged the Russian port of Sevastopol, receiving their supplies from the nearby harbour of Balaclava. In order to sever this supply route and disrupt the siege, in October 1854 the Russian commander, Prince Menschikov, attacked the Turkish redoubts along a feature known as the Woronzov, or Causeway, Heights, in the hope of being able to press on towards Balaclava itself. After several hours of fighting, the Turkish positions were taken and Russian cavalry began to advance from the Heights. The infantrymen of the 93rd Highland Regiment, led by Sir Colin Campbell, had formed a 'thin red line' just above the port, and their fusillade, along with the fire of a Turkish battalion and a light battery, checked the Russian horsemen. However, a second body of around 2,000 Russian cavalry had begun its descent from the Heights at the same moment and there were no troops available to halt them.

At that moment, the British Heavy Brigade was repositioning to cover the approaches to Balaclava. Led by the formidable Sir James Scarlett, the 900 mounted men of the Royal Scots Greys (2nd Dragoons), the 6th Inniskilling Dragoons and 5th Dragoon Guards, were immediately brought into line to face this threat. The Russian cavalry continued to advance. Scarlett did not hesitate, despite the fact that he was heavily outnumbered and would have to advance uphill across broken terrain. He gave his orders with precision and his men responded accordingly – the charge was sounded, and the mass of British horsemen started to trot through the dismembered boughs and stumps of an old orchard and a vineyard.

Characteristically for a mid-Victorian cavalry regiment, more attention seemed to be paid to the dressing (alignment) of the ranks than the enemy, and the latter were held in some contempt. As if unsure how to respond to the British approach, the Russian horsemen halted – it seemed impossible that the smaller British force would attempt to charge. The Heavies, if they kept going, would have the advantage of a shock effect against a stationary formation.

To the astonishment of the Russians (and allied observers on hills nearby) the Heavies simply ploughed into their ranks. Scarlett, who was some yards ahead of his men, disappeared as his first line of horsemen seemed momentarily to have been enveloped by the grey mass of the Russians. Allied observers on the heights a few hundred yards away echoed the sentiments of Lord George Paget who whispered: 'How can such a handful survive, much less make headway against such a legion? They seem surrounded and must be annihilated: one hardly dares to breathe.'[8]

Survivors later explained that they had cut and thrust like demons once they were in the midst of the Russian ranks. Eyewitnesses also referred to a roaring sound – the combination of hacking, clattering of blades and cursing by both sides. Lieutenant Elliot, Scarlett's Aide de Camp, cut about him, and his horse, panicked by the press of other animals around him, kicked and lashed out. A Russian trooper struck him in the forehead, causing a deep wound; a second Russian also slashed his face. As he was carried deeper into the Russian mass, he was wounded a total of 14 times, but survived, to be categorized only as 'slightly wounded'.

The Russian troopers wore thick, felt-lined jackets, which only the most furious sword blows could penetrate, giving them some protection.[9] On the other hand, the Heavy Brigade knew that its survival depended on every man

Balaclava harbour, here in a photograph by Roger Fenton, was the crucial allied base and the objective of the Russian army.

fighting to the last. Many troopers sustained multiple wounds but continued to battle on. A second line of horsemen crashed into the Russians. The momentum of their attack, such as it was, disrupted the Russian formation and, as the engagement continued, the Russians concluded that they could not overwhelm the British. As often in battle, the psychological effect was as great as the physical: after just eight minutes, the Russians began to give ground. The last squadrons of the brigade burst on to the disordered Russian flanks with such force that some Russian horsemen were thrown from their mounts. The rear ranks of the Russians began to peel away, and soon the whole formation retreated with them.

The Heavies pursued them up the slopes of the Causeway Heights and then, exhausted, and with their horses blown, they halted. Observers nearby threw their hats in the air and cheered. Sir Colin Campbell of the 93rd galloped over to give a personal and emotional congratulation. Tragically, Lord Raglan decided to issue a hasty instruction to the Light Brigade waiting below in an adjacent valley to 'advance rapidly to the front, and try to prevent the enemy

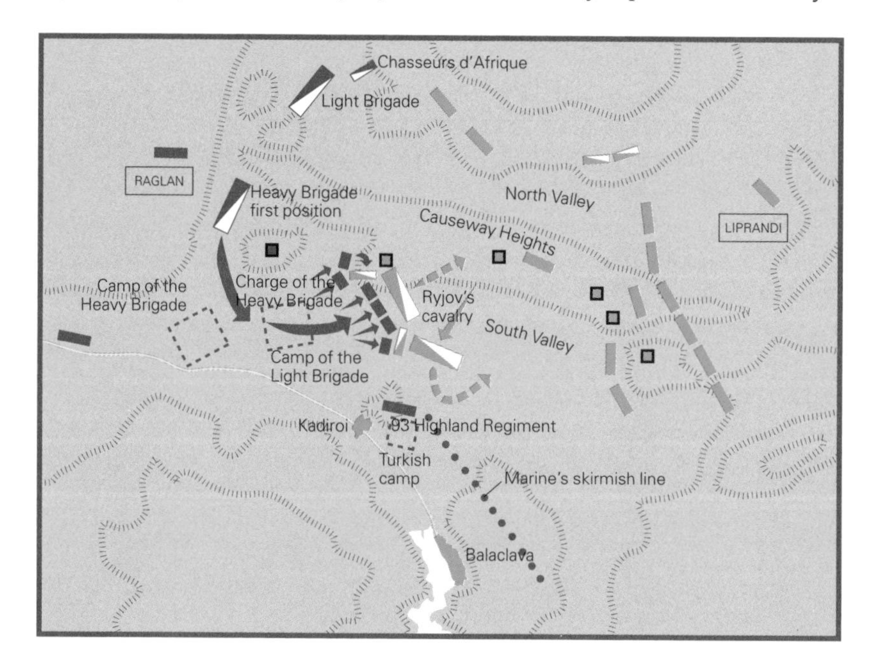

The Charge of the Heavy Brigade at Balaclava exemplified the impact of 'shock action'.

from carrying away the [Turkish] guns'. He failed to specify that the guns in question were on the Causeway Heights and the confusion resulted in the charge by the Light Brigade into the teeth of Russian batteries situated at the end of the valley. Instead of capitalizing on the success of the Heavy Brigade, the allies made an error which resulted in appalling casualties. Nevertheless, while this disaster has often overshadowed the success of the Heavy Brigade, it was the actions of Campbell and Scarlett, particularly the Heavies' shock charge, which saved the harbour at Balaclava, and, ultimately, along with the Russian commanders' timidity, determined the outcome of the battle.[10]

Analysis

Battle can often be won by an abrupt intensification of the struggle, usually accompanied by some degree of surprise, and perhaps marked by the entry of an element of the army hitherto not deployed. At both Arsuf and Balaclava, heavy cavalry provided the 'shock action' that broke their opponents' forces. Cavalry could act as a mailed fist to punch into enemy lines, combining mobility and momentum even if many knee-to-knee 'charges' were delivered at little faster than a trot so as to maintain cohesion. The charge also had a psychological effect on the enemy. It took considerable courage and discipline for foot soldiers or light cavalry to stand fast against a line of charging horses. The combination of good timing and cohesion gave the charges at Arsuf and Balaclava their devastating effect.

There are more modern examples where shock action has been provided by intense fire, rather than the physical momentum of men or horses. Air strikes and armoured thrusts have superseded the heavy cavalrymen; short and intensive artillery bombardments, and even devastating munitions (conventional or nuclear), provide a similar shock effect. Yet, despite these changes in technology, the principles remain the same.

12

Cerignola, 1503
The Hindenburg Line, 1918

Co-ordination of Fire and Movement

In manoeuvring to defeat an enemy army, timing can be critical. The arrival of troops at a precise point and time can make the difference between victory and defeat. There is often a 'moment of decision', or a 'tipping point', from which the outcome of a battle is clear. Conversely, failure to co-ordinate an attack with the fire of supporting assets such as artillery can spell disaster. There are many examples of troops running into their own gunfire, a problem that has been exacerbated by the increase in indirect fire and greater ranges of weapons systems since the beginning of the 20th century. Co-ordinating fire and movement is a fundamental principle practised in modern armies right down to squad level, but is just as important at the higher operational level. Air strikes, the movement of armour and infantry, or the precise moment for the deployment of bridging equipment might be examples in conventional warfare operations, but the principles also apply in counter-insurgency operations (see **25**), where the ability to co-ordinate civilian agencies, provide protection at the right place and time, switch quickly from riot control to more conventional war fighting, or bring in surveillance equipment are all relevant organizational factors.

At the Battle of Cerignola in 1503 the combination of firepower, entrenchments and cavalry was effective in defence.

Cerignola, 1503

The Spanish general Gonzalo de Córdoba won a significant victory at Cerignola in southern Italy in 1503 against a much more numerous French army using combined pikemen and arquebusiers. Protected by field defences, Córdoba's hedge of pikes warded off two French cavalry charges, and his cannon broke up their formation. A third French attack fell on his right flank, but the fire of massed arquebuses and the wall of pikes was sufficient to blunt the assault. A final French attack came from mercenary Swiss pikemen, but they too were thrown back. As they retired, the Spaniards moved in pursuit, with arquebusiers and pikemen offering mutual support. This combination of 'fire and movement' (co-ordinated movements covered by fire), and the mutual support of infantry, artillery and cavalry was a formula which allowed Córdoba's 8,000 to defeat 32,000 French and Swiss.

The co-ordination of all arms, in the more complex environment of 20th-century warfare, is exemplified by the battles to break the final German defence lines in the First World War in 1918, with armour now replacing cavalry.

The Hindenburg Line, 1918

The 'Hundred Days' campaign on the Western Front in 1918 was the most successful offensive conducted by the Allies in the First World War. In contrast to the trench stalemate of the early years, this was a battle of movement and incremental achievements. It has been overshadowed in subsequent perceptions, not least because many sought to portray the whole conflict as a futile exercise which produced no clear victor.[1] Yet from a tactical and operational point of view, it was an undeniable success.[2] Reflecting on the fighting on 8 August 1918, the German Chief of Staff,

Previous pages: Battlefield victory requires the careful co-ordination of units and logistics; here infantry in France in 1918 advance by 'fire and manoeuvre'.

Erich von Ludendorff, remarked that his defeats marked 'the Black Day of the German Army' which convinced him that Germany could no longer win the war.[3]

This outcome was all the more remarkable because Ludendorff had anticipated a military triumph in the spring of 1918. He had redeployed large numbers of men from the Eastern Front because Russia had withdrawn from the war, and he believed this numerical advantage would enable him to overrun the Western Allies before the Americans arrived in force. Furthermore, he was aware that France had suffered a haemorrhage of manpower in the Battle of Verdun (see p. 202) and also that the British were extended over a long front. However, the British and French managed to absorb the shock of the German spring offensive and by midsummer they were ready to turn to the offensive again. One problem the Allies had to overcome was co-ordinating their various national forces. The Americans and the Dominions of Canada and Australia insisted that their troops fought as composite formations, refusing to see their units spread among British or French corps. The decision to create a supreme Allied command largely settled this question.[4]

The Allies' success derived both from this ability to co-ordinate the various arms and to apply the military lessons learnt in previous years. Instead of the laborious bombardments of 1915 and 1916, the Allies opted for surprise. A deception plan gave the impression that the Canadian Corps had been moved to the Ypres sector. Reinforcements brought in from the Middle Eastern and Mediterranean campaigns were deployed without the Germans being aware of their arrival. Local attacks were authorized near St Mihiel and Ypres to suggest that the Allies were preparing major attacks there. Movements in the Amiens sector were restricted to darkness, and daylight movements were invariably feints. Strict security was maintained among all the troops and this was sustained even when a local German counter-attack by the 27th Division (a storm troop formation) seized some Allied trenches earmarked as a jumping off point for the offensive. No attempt was made to regain these trenches so the Germans would not be tempted to reinforce the area.[5]

12

The breaking of the Hindenburg Line in 1918 demanded the orchestration of infantry, armour, artillery and air power.

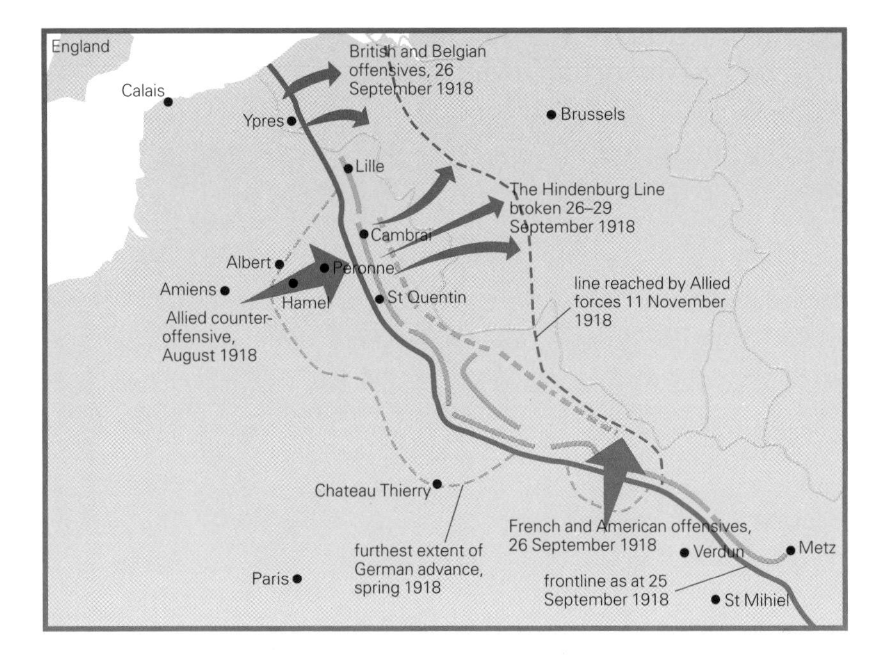

On the day of the offensive, 8 August, thick fog blanketed the battlefield. At 0420 hrs, large formations of tanks – 580 on the first day of the Battle of Amiens – rolled forwards ahead of the infantry, and a relatively short barrage was delivered only when the other arms had actually left their start lines. The attack was spearheaded by the crack Australian and Canadian Corps. Co-ordinated artillery fire from 1,386 guns laid down a creeping barrage in short, intensive bombardments which were 'lifted' a few hundred yards before the infantry attack was delivered. Thus the troops went in just behind a curtain of exploding shells. At Hamel, the Australians overran the German defenders even though they were dug in deeply and could sweep the approaches with machine gun fire. Indeed, the Germans seemed to be taken by surprise and reacted relatively slowly to the commencement of the attack.

With close air support from the RAF, artillery fire could be adjusted from map calculations and air photographs and directed from the air as the action

Infantry advance through bursting shells in the exposed landscape of the Western Front in 1918.

unfolded. German positions were also strafed by Allied aircraft. At 0730 hrs, the advancing troops were 4,000 yd (3,660 m) into the German defences. Some troops were delivered through gaps that had been created by specially designed infantry-carrying tanks, while cavalry rode alongside, eager to push on into the German rear. By the end of that day, the German positions had been broken open to create a gap 15 miles (24 km) wide through which cavalry and armour, including faster light tanks, poured. These drove deep into the German rear, causing panic. The Allies took 17,000 prisoners and 330 guns, inflicting 30,000 casualties for the loss of 6,500.

As the momentum of the Amiens attack began to falter because of the difficulties of keeping forward units supplied, Field Marshal Douglas Haig, the British commander on the Western Front, launched a second offensive on the Somme at Albert, 12 miles (19 km) north of Amiens, on 21 August. It was another success, with the First Army capitalizing on the achievements of the Second Army as they leapfrogged through the German defences. The first attack pushed the Germans back on a front of 34 miles (55 km) and this was widened by a further 7 miles (11 km) just four days later. Between 31 August and 4 September, the Australians once again led the assault. They crossed the Somme and broke the Germans at Mont St Quentin and Peronne.[6] By the middle of September, the German salients in the sector had been pinched out. To stabilize the situation on the Western Front, Ludendorff was now dependent on a line of deep fortifications known as the Hindenburg Line. However, many of his troops were showing signs of fatigue and low morale.

Lieutenant General Sir John Monash, the commander of the Australian Corps, described the co-ordination of the Allied offensive as: 'a score for a musical composition, where the various arms and units are the instruments, and the tasks they perform are their respective musical phrases. Each individual unit must make its entry precisely at the proper moment, and play its phrase in the general harmony.'[7] Captain D. V. Kelly, of the 6th Leicesters, reflected in similar vein: 'By September 1918 ... [it] is very important to remember that the artillery had improved their technique just as had the staffs and the infantry: in 1916 one could hardly have relied on the accuracy and exact synchronization, which now one had learned to expect.' The historian Gary Sheffield concluded: 'Co-operation between the various arms was the key to the BEF's [British Expeditionary Force's] victories in 1918.'[8]

On 26 September, Marshal Foch, the Supreme Allied Commander, launched attacks at either end of the Front – the British and Belgians in the Ypres sector and the Americans at Meuse-Argonne – which made some initial headway. Shortly after, the Canadians penetrated the Hindenburg Line at Cambrai. The main attacks came in the centre: the Australians and United States II Corps made some progress, while the British Fourth Army, containing a number of experienced divisions, broke into and then through the Hindenburg Line.

The 46th (North Midland) Division of the British Fourth Army was ordered to cross the St Quentin Canal, which ran like a moat in front of the Hindenburg defences. The banks were in places quite precipitous and laced with barbed wire and concrete gun emplacements, while the bottom was filled 8 ft (2.4 m) deep with water or glutinous mud. The general opinion among the 'Other Ranks' was that the mission was doomed to failure, and they suspected they were little more than a diversion to cover an attack further north where the canal entered the 3-mile (4.8-km) long Bellicourt Tunnel. However, in the days before the attack, the infantry managed to overwhelm each of the German positions by advancing in loose platoon-sized groups, in conjunction with tanks, preceded by an artillery barrage and with close air support.

At dawn on 29 September the Division attacked the canal. It was misty, and indirect massed machine gun fire poured into the gloom. The leading infantrymen overran the first German line, and found the trenches full of enemy dead. Small groups plunged across the canal, and one party found itself in the shell holes in front of the Riqueval Bridge, which was, miraculously, intact. Without hesitation they dashed forward, half expecting the bridge to be blown up beneath their feet. Within seconds they were across, and this bridgehead proved invaluable in getting artillery across the canal far more rapidly than expected. The Division pressed on, deeper into the German defences. One officer remarked:

12

> Suddenly the mist rose and the sun of our 'Austerlitz' appeared, strong and refulgent. Over the brow of the rise opposite to us came a great grey column [of prisoners] ... and then another, and another and another. We had broken the Hindenburg Line, and 4,200 prisoners, seventy cannon, and more than 1,000 machine-guns were the trophies of the fight gathered by our single division![9]

The Germans were now pressed back across the entire front. Within weeks, they were practically at the German border and their logistical system had collapsed. While some units showed defiance to the very end of the war, large numbers were captured or retreated in disorder.

Within days, Crown Prince Rupprecht reported the imminent collapse of the German army and believed that no effective resistance could be sustained. As a result, the German authorities agreed an armistice on 11 November 1918, bringing the war to an end.

British tanks, seen here at Bellincourt, in September 1918 with trench-crossing fascines, also known as 'cribs', punched through the German defences with the close support of Allied infantry.

Analysis

The concept of co-ordination is closely linked to that of concentration (see **13**), and emphasizes the right combinations of force structures and timing. The appropriate weight of fire, numbers and types of formations, logistical support and specialist equipment must all arrive at precisely the right time and location for their task. This is a demanding challenge. General Córdoba managed to ensure his pikemen and arquebusiers offered each other mutual support as they advanced at Cerignola, and their movement was supported closely by cavalry and artillery. On the Hindenburg Line, the outstanding features were the volume of fire (with deception shelling, short bombardments for surprise and curtains of fire dropped in front of the advancing troops), new flexible infantry tactics that stressed dispersal and devolved command, air supremacy, good communications and the close co-operation of all arms. Clearly, on this vast scale, co-ordination of all the arms and services was a significant achievement and was instrumental in producing victory.

13

Jagdgeschwader, the Western Front, 1916–17
Midway, 1942

Concentration and Culmination of Force

The concentration of force was regarded by Clausewitz as the first and highest principle of war.[1] He reiterated that, at the operational level, commanders had to concentrate maximum force, which in his day equated to all available troops, at the decisive point; it was essential to overwhelm and break the enemy physically and morally. Commanders should avoid detaching troops unless absolutely necessary, so as to retain the maximum combat power in one place. He also stressed the need to ensure that troops should not be allowed to be idle or distracted by subsidiary operations, a concept he referred to as an economy of force. He posited that the concentrated force of an army should be unleashed at the centre of gravity – the critical location or moment in a battle (see **1**). A commander first had to identify accurately where and when the centre of gravity, or, as Baron de Jomini put it, the 'decisive point', actually lay. Only then could the relevant forces be marshalled into position.

Previous pages: US dive bombers on their way to attack the Japanese carriers at Midway.

To strike the enemy at the decisive point on the battlefield requires the accumulation of troops, weapon systems, logistics and other assets (such as siege weapons, bridging equipment or transport). On campaign, this might demand that a great complexity of formations and sub-units are brought together at the right time and place, requiring efficient staff work (see also **12**). The difficulty for armies is that casualties, disease, the need to man depots or occupy territory, and a stretched logistical supply system can reduce the speed and power of an offensive. Clausewitz referred to this as the culminating point of a campaign. Beyond a certain point, always difficult to identify, an attacking force may be weakened, and should consider going over to the defensive until its strength is restored. For success, armies must ensure that they bring the enemy to a decisive engagement before the culminating point has been reached. Moreover, they must make certain that they draw together and concentrate their force at that decisive point.

Nevertheless, it is also true to say that a stubborn refusal to disperse any troops, and thus miss the opportunity to extend the enemy's line, actually negates to some extent the advantage of concentrating one's own forces. Indeed, the purpose of concentrating is to achieve a *local* superiority of numbers, not necessarily an absolute numerical superiority. In other words, the numbers brought to the decisive point can, in fact, be quite small. In modern conventional operations, the question of numbers is less important than the firepower the force possesses and which it can command from supporting assets. Thus, it is fire that must be concentrated at the decisive point. In the two world wars, that fire was often manifest in air power, and two examples are presented here to illustrate how victory through such concentration was achieved.

Jagdgeschwader, the Western Front, 1916–17

In the first months of the First World War, the role of aircraft was restricted to reconnaissance, and few envisioned that this conflict would shape air warfare tactics in the future. Reconnaissance from the air was particularly important because it was impossible for cavalry to carry out this duty without coming under effective fire. Air photography and systems of artillery spotting soon evolved, and, as early as 1915, pilots were also making attempts to shoot down their opponents' reconnaissance flights. Machine guns were fitted to two-seater aircraft as protection for observer teams, but the development of interrupter gearing (which allowed machine guns to fire forward through the propellers) permitted the creation of specific 'fighter' planes.[2] In mid-1915, the Germans began to assert their superiority with this technology on the Western Front, but the Allies quickly introduced new aircraft of their own. For most of 1916 the Allies held the upper hand, and their offensives of that year, at Verdun and the Somme, profited from the air superiority they enjoyed. However, in late 1916 the Germans began to wrest control of the skies back. In 'Bloody April' 1917, the Allies were losing pilots at a rate of 30 per cent a week – a result of continuous offensive patrolling, reconnaissance and sorties. In the race for air supremacy, attrition, but also aircraft design, became important.

Tactical developments meant that flying in groups of six aircraft became more common than individual flights. These groups could either act offensively, seeking out targets in the air or on the ground, or provide protection for slower bombers or observer planes. In defence, aircraft were deployed in layers, reflecting their characteristics and suitability to operate at certain altitudes. The German strategy of using their squadrons more defensively in 1917 enabled them to marshal their reserves, strike only where they were needed, prioritize their resources, and preserve the lives and therefore the experience of their pilots.

To counter Allied sorties at strategic points on the front, groups of four individual *jastas* ('squadrons') were assembled into larger units, known as Jagdgeschwader ('hunter squadron', or fighter wing) formations, sometimes numbering more than 30 planes. Separate *jastas* could be concentrated quickly to meet a particular threat as it emerged,[3] and certain *jastas* made up of experienced pilots would be transported from different parts of the front by rail, with all their aircraft and logistical support, and deployed in a critical sector. This concentration of force was highly effective: local air superiority could be restored, or at least Allied supremacy contested and their operations disrupted, thus enhancing the strategic defence of ground positions below.

Experienced pilots who survived the high attrition rates could, particularly when grouped together in this way, develop and exchange ideas about aerial tactics. Some techniques emerged in the course of flying and fighting continuously for months, but improved engines and aircraft also made it possible to evolve new approaches: aircraft became specialized – for speed, rate of climb, manoeuvrability or armament. Scout aircraft, for example, exploited their speed and manoeuvrability to escape destruction, and some used a spinning nose dive to evade their attackers; this technique had been almost impossible before 1916 because it would have resulted in a loss of control of the aircraft. French pilots pioneered the 'rolling' technique along a longitudinal axis; and the 'Immelmann turn', a long looping movement while gaining height, which made it possible to make sudden turns, was invented by a German pilot of that name. Lone pilots developed the technique of flying 'out of the sun', using the blinding effect to conceal an approach. Clouds could also provide useful cover.

If attacked suddenly, pilots would generally bank and dive. However, this sometimes played into the hands of attacking aircraft which operated in pairs. As one flew broadside, and opened fire at a distance, the target aircraft would seek to evade it but then be attacked from the rear by the second plane. If the target aircraft turned to fight the initial attack, the second attacking plane would also turn swiftly to come in from behind. A burst of gunfire lasted a few seconds at best so a sudden attack or the manoeuvre prior to the use of machine guns was critical to the outcome. Again, aircraft design and function could make a difference. Two-seater aircraft were often slower but they protected themselves with a rear-mounted machine gun. Fighter planes did not always manage to defeat the two-seaters, and a combination of fighters and two-seaters, grouped *en masse*, was a formidable arrangement. If groups of aircraft could be concentrated quickly, and their formations co-ordinated, they could overwhelm an adversary.

One of the most accomplished pilots of the First World War was the 'Red Baron', Manfred von Richthofen. Despite his formidable personal achievement of 80 confirmed 'kills', it was his fighter group, Jagdgeschwader

1, or JG-1, that caused considerable damage to Allied pilots. JG-1 decorated its aircraft in bright colours to assist identification in dogfights. The distinctive red of Richthofen's own plane (in fact, he flew several types of aircraft) led to his famous nickname, although when he was in command of Jasta 11, prior to leading the entire group, almost all his pilots adopted red as their colour. These bright colours and the resemblance to the transport and rapid redeployment of circus groups earned them the name the 'Flying Circus'.

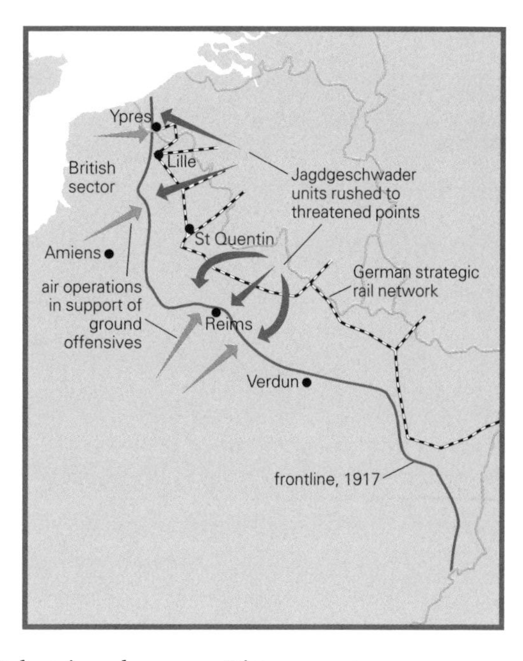

The rapid movement and concentration of German squadrons over the Western Front using the rail network ensured local air superiority.

Despite the propaganda image of the lone Red Baron downing aircraft in single combat, Richthofen was a master of concentrating force and teamwork. He insisted on a set of flight doctrines, known as *Dicta Boelcke* after his former colleague, Oswald Boelcke, who flew with him on the Eastern Front, to ensure that individual pilots offered mutual support and

An Albatross D.II, one of the German fighter planes that equipped the *jastas* as they attempted to gain control of the air over the Western Front.

Manfred von Richthofen, nick-named the 'Red Baron', was the scourge of Allied pilots. Although he had a reputation for acting independently, in reality he worked in concert with other members of his squadron.

acted within the mission objectives of the whole squadron. Richthofen would attack from the sun, for example, while other pilots maintained a watch over his flanks and rear. He did not embrace sophisticated aerobatics, preferring to manoeuvre his group to concentrate as much firepower as possible on the selected targets. Contrary to the flamboyant image of fighter pilots, Richthofen was rather austere and distant. Other accomplished pilots were also part of experienced teams that had learnt to offer mutual support in combat, and they increased the numbers of 'kills' partly by being able to survive for longer in this way. However, although air 'aces' attracted attention, and were exploited for their propaganda potential, it was the tactical, organizational and technological changes that were of far greater significance in the longer term in the air war, and the ability to combine at the decisive point was a critical factor.[4]

Midway, 1942

The American victory at the Battle of Midway provides an excellent example of concentration and culmination. The Japanese intention was to lure the American aircraft carriers that they had failed to destroy at Pearl Harbor in 1941 and the Battle of the Coral Sea in early 1942 into a trap. They calculated that a movement towards Midway Island would compel the Americans to protect the area, as it lay on the approaches to the strategically important Hawaii group. To achieve surprise, the Japanese navy was divided into a carrier strike force and a group of warships, the latter moving several hundred miles behind the carriers. Having drawn the Americans into action, the Japanese carrier aircraft would destroy the US carriers, whereupon the surface ships would deal with the vulnerable American fleet and any shore defences.[5]

Admiral Chester Nimitz, Commander in Chief of the Pacific Ocean Areas, was alive to this Japanese threat – Japanese codes had been broken – and he prepared accordingly. Every effort was made to repair USS *Yorktown* to give the Americans three operational carriers against the Japanese four. Task Force 16 consisted of the carriers *Enterprise* and *Hornet*, with six cruisers and nine destroyers; Task Force 17 was made up of *Yorktown*, with two cruisers and five destroyers. The island of Midway itself provided an additional space from which American fighters and B-17 bombers could be launched. On 3 June 1942, nine American bombers located and attacked the Japanese transport ships, confirming intelligence reports of an imminent strike against Midway; by contrast the Japanese were still struggling to obtain accurate information about American deployments since their submarine and air reconnaissance were inadequate. As a formation of American bombers left Midway, the first Japanese air element from the carrier group was nearing the island. Japanese aircraft were able to shoot down a number of US fighters around the island, but they took some significant losses from anti-aircraft guns. Ten American bombers made an attack on the Japanese carriers, but were repelled.

The American carriers were now approaching rapidly. While the Japanese were preparing to launch a second wave against Midway, they were hit by successive lines of American dive bombers, torpedo planes and fighters.

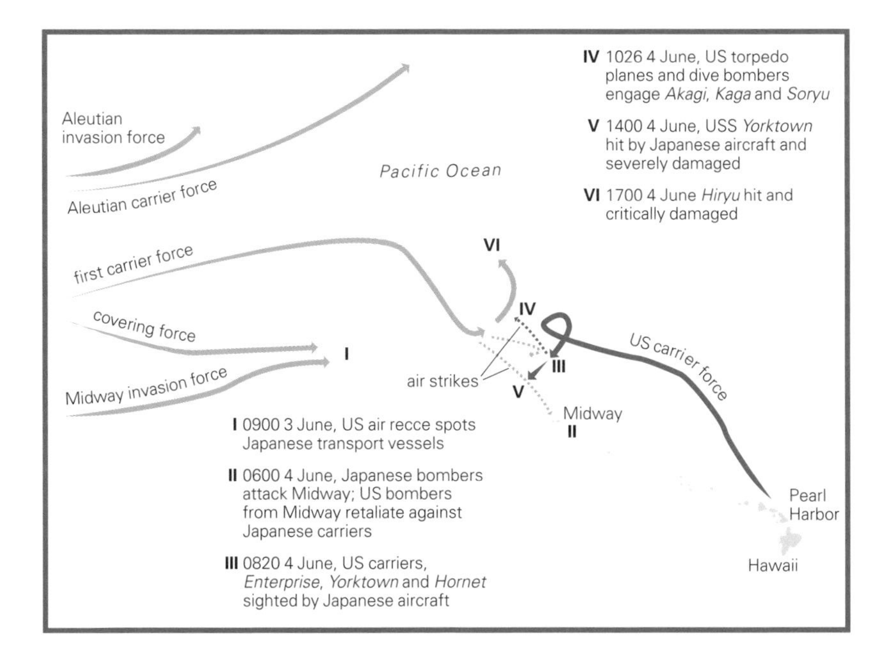

The Battle of Midway was decided by the concentration of American forces at the critical point and time.

Pilots had to endure a hail of anti-aircraft fire to reach their targets, but co-ordinated attacks could devastate an enemy fleet. Here USS *Yorktown* is under attack during the Battle of Midway.

However, of the 41 slow-moving torpedo planes, 35 were shot down. The imperative of launching the attack had prevented concentration into a single strike, leading to heavy losses as each attack became the focus of intense Japanese fire. Yet the waves drew the Japanese fighter escorts into action, mainly southeast of the fleet's location. Thirty-seven American dive bombers from the carrier *Enterprise*, approaching from the northeast and southwest, were therefore able to close with the Japanese vessels. Three of the Japanese carriers were attacked by aircraft swooping with total surprise from 19,000 ft (5,800 m). Within six minutes, they were crippled. On the *Akagi*, one bomb penetrated the upper deck and exploded among armed and fuelled aircraft in the hangar below, causing devastation. *Soryu* and *Kaga* were also hit several times in the hangar decks, which were packed with planes being prepared for a second strike against Midway. Fires raged throughout the carriers and it was clear that they would play no further part in the battle.

As the Japanese aircraft were refuelling, vulnerable on the decks, one Japanese pilot remarked:

> I looked up to see three black enemy planes plummeting towards our ship [*Akagi*]. Some of our machine guns managed to fire a few frantic bursts at them, but it was too late. The plump silhouettes of the Dauntless dive-bombers grew larger, and then a number of black objects suddenly floated eerily from their wings – Bombs! Down they came straight towards me! I fell intuitively to the deck ... The terrifying scream of the dive bombers reached me first, followed by the crashing explosion of a direct hit. There was a blinding flash and then a second explosion much larger than the

> first. I was shaken by a weird blast of warm air. There was still another shock ... apparently a near miss. Then followed a startling quiet as the barking of the guns suddenly ceased. I got up and looked at the sky. The enemy planes were already long gone from sight.[6]

Yorktown's dive bombers arrived soon after the first attack and added to the carnage. *Soryu* was so badly damaged that it was abandoned just 20 minutes after being hit.[7]

The remaining Japanese carrier, *Hiryu*, made two counter-attacks against *Yorktown*, which was badly damaged. However, American aircraft were now hunting this final Japanese carrier: in mid-afternoon they found it and another dive bomber wave set the *Hiryu* ablaze. It sank at 0500 hrs the next morning. The Americans made further attacks against the powerful Japanese surface fleet in the following days, but were careful not to risk their own surface forces, keeping their carriers out of range. The Japanese withdrew: it was clear that this battle had given the Americans the initiative in the Pacific.

Analysis

Both the examples of air concentration, on the Western Front and in the Pacific, illustrate the applicability of Clausewitz's maxims in land, air and naval warfare. At Midway, the American air attacks were not completely co-ordinated, but sufficient force had been concentrated by Admiral Nimitz in advance to overcome this tactical problem. On the Western Front, the Germans were able to rush squadrons to threatened points, concentrating them to deal with specific offensives while preserving their resources as far as possible. Such local superiority, even if temporary, has often proved far more valuable than the dissipation of forces in an attempt to be strong everywhere.

14

Eben Emael, 1940
Pegasus Bridge, 1944

Seizing and Retaining the Initiative

It is a fundamental principle that commanders seize and retain the initiative in battle. Keeping ahead of the enemy's decision-making – by reading the situation, issuing orders and effecting changes before the enemy can react – can enable a force to dominate a battle or a campaign. Similarly, increasing the tempo of operations and forcing the enemy to respond to a rapid series of manoeuvres can have the same effect. An alternative is to make unexpected and sudden challenges, seizing strategically important points such as a bridge, mountain pass or castle by surprise in a *coup de main*, or to carry out deception plans.

Taking the initiative is essential, but retaining it is just as important. If the enemy is given respite, they can reorganize, mount an effective defence or counter-attack. By contrast, maintaining momentum in the attack can cause considerable disruption to enemy formations and damage enemy morale significantly. It is a military maxim that once an enemy is on the run a commander should keep them running. By these means, even relatively small forces have been able to defeat larger ones.

14

Previous pages: American paratroopers landing in Normandy, France, as part of the D-Day operations. Airborne forces have the ability to leapfrog the enemy's lines and take vital ground, thereby seizing the tactical initiative from an adversary.

Above all, the enemy should be compelled to 'dance to your tune', since an army or fleet that acts with the greatest speed and resolution wins the day. In Genghis Khan's campaign against Muhammad II, Shah of Khwarizm in Central Asia, in 1219, the Mongol cavalry used their speed to seize the initiative, and changed the axes of their attack to retain it.[1] The Mongols had prepared their offensive carefully in the face of a stronger army. After a small demonstration in the Ferghana Valley, which drew a significant proportion of Shah Muhammad's army to the south of his realm, several months elapsed. A Mongol force then surprised and destroyed a settlement in the far north of the Shah's domains, sucking in Khwarizm reinforcements. The Mongols now increased the tempo: a strong detachment attacked outposts and forts along the Syr Darya River, few of which could offer mutual support. At the same time, a second strong detachment of the main army, under General Jebe, swept around the rear of the Shah's army. When the Shah's relatively slow-moving army turned to engage Jebe, there were too few troops to deal with Genghis Khan's main body which had moved at great speed to appear in front of Bukhara to the south. The Shah's forces simply could not react fast enough. Bukhara and Samarkand fell in quick succession, the Khwarizm army began to break up and the Shah was forced to flee.

The Mongol and Chinese military concept of 'slow-slow: quick-quick' had been applied against Muhammad II's forces. Slow preparation and a long delay – the second 'slow' – between the initial feint in the Ferghana Valley and the subsequent attacks had given the Mongols plenty of time to formulate their offensive plan. Muhammad II failed to seize the initiative and remained inactive on the defensive. When the attack came, it was sudden and from an unexpected direction (the first 'quick'). The Mongols increased the tempo and moved fast to prevent any recovery (the second 'quick'); changing horses on the march (each rider had three mounts) and living from mare's milk and the food they carried in their saddle bags, they barely paused in their advance. As the Shah's forces lumbered into new positions to meet the threat, the Mongols had already changed direction. The rapid arrival of the main body, its movements previously concealed by the fast-moving operations elsewhere, was sufficient to break the Khwarizm forces.

Napoleon's armies in southern and central Europe earned a reputation for surprising their enemies through speed of operations. Foraging for rations every few days, the French soldiers carried biscuits, up to 80 rounds of ammunition and only basic utensils in their knapsacks; they were not burdened with the scale of slow-moving baggage wagons that encumbered most armies.[2] Napoleon exploited the fact that his troops could march 20 miles (32 km) per day, to give the impression of being able to appear anywhere, including behind his enemies – the so-called *manoeuvre derrière*.

As late as 1815 in the Waterloo campaign, Napoleon 'stole a march' on the allies and advanced so rapidly into the Netherlands that his adversary, the Duke of Wellington, barely had time to concentrate his dispersed formations. Indeed, it was with the utmost difficulty that Wellington's army, fighting a 'meeting engagement' (that is, deploying from the line of march straight into action), was able to hold, briefly, the strategically important crossroads at Quatre Bras. Wellington's original idea of concentrating the Prussian and Anglo-Dutch-German armies had to be temporarily abandoned. Reflecting on his enemy's speed of manoeuvre, Wellington admitted that Napoleon had 'humbugged me'.[3] Napoleon failed to capitalize on this initial advantage, however, and Wellington's ability to delay the French stole the initiative and gave the Prussians the opportunity to converge on Napoleon's flank at Waterloo and defeat him.

Covert operations and raids have sometimes fulfilled the criteria of seizing the initiative, although many commanders have dismissed Special Forces' activities as an expensive waste of resources for limited results. Clausewitz was critical of deception plans, diversionary actions and 'irregular' operations. He believed they failed to achieve much, and, worse, distracted commanders from the crucial principles of concentrating maximum force at the enemy's centre of gravity.[4] Nevertheless, there are historical examples of successful minor operations. In North Africa in the 1840s, French forces conducted *razzia* (raiding) attacks against their Algerian opponents. In 1876, the United States army was able to disrupt the Native American tribes by raiding their camps in winter, following General George Armstrong Custer's defeat at the Little

Bighorn. At Slim Buttes in South Dakota, in September, Captain Anson Mills and 150 men made a dawn attack, scattering the villagers of 37 lodges. In late November, Colonel Ranald Mackenzie made a raid with 1,000 troopers on the Dull Knife Cheyennes. The destruction of their lodges and ponies in freezing weather forced many to surrender.[5]

Armies have, on occasion, 'taken the war to the enemy's homeland' to regain the initiative. Forcing an enemy to react to unexpected attacks against their logistics, command and control systems, or the very fabric of their society, can completely alter their strategy or ability to fight effectively on the battlefield. Cutting the enemy's line of communication, the route for supplies and information, could quickly neutralize a campaign or battlefield adversary. Air interdiction, usually through strategic or tactical bombing, and the destruction of logistics in depth was frequently used in the Second World War to deprive opponents of reinforcements and supplies.[6] During a conflict, a scorched earth policy, blockade or longer term denial of resources might regain the initiative by forcing an enemy to react, or to change strategy.

Eben Emael, 1940

In modern warfare the *coup de main* has often involved aerial outflanking with parachutists or air mobile forces. In 1940, at the outbreak of hostilities in the Second World War, German airborne troops of the 7th Flieger Division were tasked with the capture of a key Belgian fort at Eben Emael, which would open the way for ground forces to speed into Belgium. The fort stood on the border, protecting three vital bridges and the strategically important 'Gap of Vise', a corridor that led to Liège.[7] The plan was audacious. They were to land with gliders on the fort's concrete roof, attack the structure with hollow charges and then assail the garrison. Ground forces were to fight their way through to support the operation and relieve the paratroopers. The alternative was to besiege the fort and its two mutually supporting fortresses, but, given that it could hold a garrison of over 1,000 men, had protective bunkers, artillery positions, retractable domes bristling with guns, and was protected by machine gun posts and anti-tank ditches, it was thought this would take weeks. The glider attack produced victory in 24 hours.

The paratroopers rehearsed their mission on forts in occupied Czechoslovakia. Careful pre-war reconnaissance had also been carried out. The landing zone for 11 gliders was tight, a triangular roof 875 by 985 yd (800 by 900 m) in size, but there were relatively few anti-aircraft defences, and most of the armament was designed for ground assault.

On 10 May 1940, the gliders began their silent descent towards the fort. They landed exactly as planned and 74 paratroopers under the command of Lieutenant Rudolf Witzig commenced their attack. These crack troops

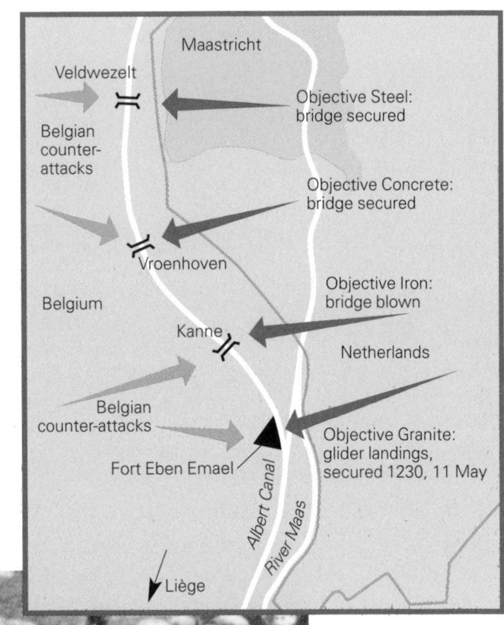

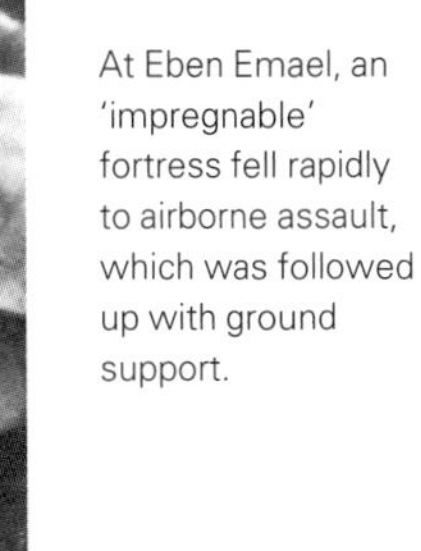
At Eben Emael, an 'impregnable' fortress fell rapidly to airborne assault, which was followed up with ground support.

German *Fallschirmjäger* (paratroopers) seized the vital territory in the Battle of Eben Emael.

were not only well drilled, they were exceptionally well motivated and prepared to take heavy casualties. The operation began as expected and had the element of surprise. High explosives were used against each of the cupolas, and a flamethrower neutralized several machine gun posts. The Belgians retaliated, trying to sweep the Germans from their precarious and exposed positions. Furthermore, they blew one of the bridges and their intense fire from the fort was able to delay the paratroopers' relief forces, until a pioneer unit managed to cross the River Maas (Meuse) with inflatable dinghies just to the north of the fort and seized the village of Kanne.

In the darkness, the Belgians put up a fierce resistance and even mounted local counter-attacks, but their firing line was effectively sandwiched between two German forces, while air attacks on their rear prevented reinforcements from coming up. Flooded terrain forced the German relief troops to confront a 65-ft (20-m) high concrete bastion on the northern point of the fortress. Once again, a small detachment of 50 men made a crossing of the flooded area in small boats, and, as dawn broke, the relieving forces and the paratroopers met up. Witzig's men continued to lay charges against the barrels of the fort's guns, and, one by one, each of the three fort complexes was neutralized. As the Belgian fire slackened, the Germans secured the two remaining bridges and brought up light artillery which was directed against the remaining forts. By 1000 hrs, the final assault was made on the last fortress and at 1230 the garrison capitulated. A thousand men were taken prisoner, many clearly shocked at the speed and suddenness of the German assault.

Pegasus Bridge, 1944

A similar bold and brilliant air operation, relieved by ground forces, was made at 'Pegasus' Bridge in Normandy in 1944 by the Oxford and Buckinghamshire Light Infantry and men of the Royal Engineers and Glider Pilot Regiment.[8] On the night of 5/6 June, 181 men under Major John Howard landed in gliders within metres of the target bridge. The aim was to secure this crossing of the Orne River and its canal to protect the D-Day landings from a German armoured counter-attack. They achieved total surprise and in just 10 minutes

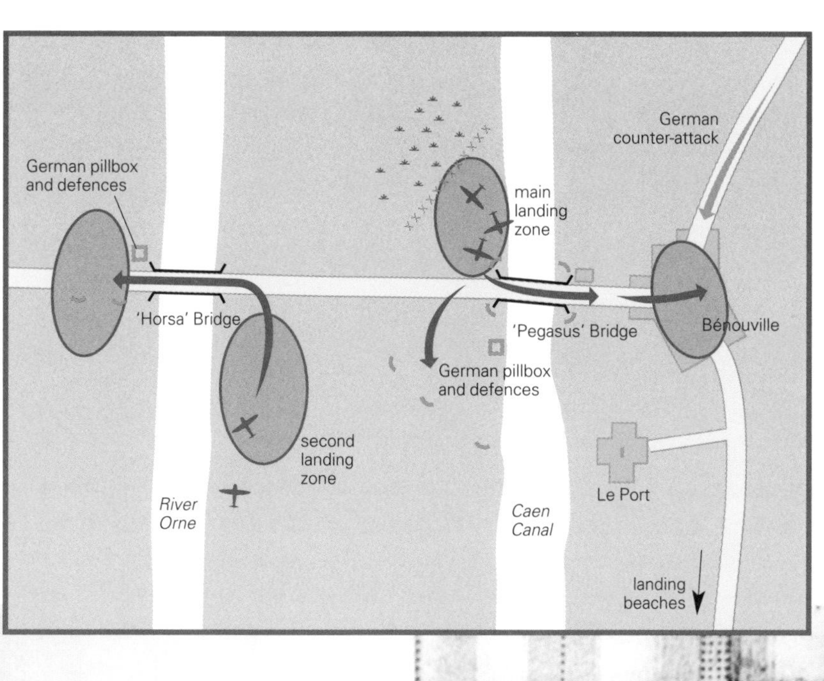

A daring and accurate Allied landing by glider secured Pegasus Bridge on D-Day, 1944.

At Pegasus Bridge ground forces relieved the lightly armed airborne units, and thus retained the initiative.

had secured the bridge. Holding the bridge against local counter-attacks, the glider-borne force was then relieved by troops who had landed at Sword Beach.

An American Waco CG-4A, used in the Second World War. Whatever their lightweight means of deployment, airborne forces can turn the tide of a campaign decisively.

Other parachute units of the brigade were scattered and found it difficult to mount operations, but the 9th Battalion, Parachute Regiment, achieved a victory despite the conditions. Only 150 out of 700 men could be mustered to attack their primary objective, the Merville Battery, whose guns threatened the landing beaches.[9] They had no artillery support or mortars and only one medium machine gun with which to assault the bunkers and casemates of a complex that was surrounded by minefields. Yet Lieutenant Colonel Terence Otway realized that delay would only worsen the situation and allow the Germans to recover, bring up reinforcements and even counter-attack. He had to seize the initiative, and ordered an attack across the minefields into the teeth of the German defences. No fewer than 65 men of his command were casualties, but their sudden attack overwhelmed the German garrison entirely. Smaller detachments managed to carry out the secondary tasks of blocking roads and cutting off the beachhead.

Analysis

The *coup de main* allows relatively small numbers of men to secure strategically important positions and exemplifies the principle of seizing the initiative. Increasing the tempo of operations can also mean that the initiative is retained: in Genghis Khan's campaign, this was certainly a contributory factor, although it should also be acknowledged that the Mongols possessed other advantages – mobility and self-sufficiency in transport and logistics, tar-dipped flaming arrows to set up smoke screens on the battlefield, terror tactics and deception. In the airborne operations in the Second World War, surprise made it possible to seize the initiative, but retaining it – and thereby ensuring success on the battlefield – required ground forces of sufficient strength and speed to support the *coup de main* troops. In everything, the element of surprise, and therefore good security prior to the operations, was essential.

At Eben Emael, the German paratroopers were equipped with enough combat power to achieve their mission. Similarly, the glider-borne troops at the Orne bridges in 1944 were the appropriate size to complete the task speedily and with surprise. A larger force might have taken longer to regroup after the landing, and decision-making slowed down as orders were passed through subordinates. All attempts to seize the initiative, including the *coup de main*, are not without risk, which must be carefully weighed against the objective. Decision-making under the pressure of war, including the decision to act quickly or boldly to gain the initiative, must therefore be made with a rational assessment of the risks and potential benefits. Clausewitz wrote: 'The higher the rank, the more necessary it is that boldness [courage] should be accompanied by a reflective mind, that it may not be a mere blind outburst of passion to no purpose.'[10]

15

Trafalgar, 1805

Off-Balancing and Pinning

Closely related to the idea of seizing and retaining the initiative (see **14**) is the method by which an enemy is pinned into a position where firepower or manoeuvre can destroy him, or where the enemy is so disrupted by off-balancing probes that he has little idea where the main attack will fall. Field Marshal Montgomery described the importance of off-balancing with characteristic clarity:

> Most opponents are at their best if they are allowed to dictate the battle; they are not so good when they are thrown off-balance by manoeuvre and are forced to react to your own movements and thrusts ... As the battle develops, the enemy will try to throw you off-balance by counter thrusts. This must never be allowed to happen. Throughout the battle area the whole force must be so well balanced and poised, and the general dispositions and layout so good, that one will never be dictated to by enemy thrusts.[1]

Montgomery understood that composing groups with the ability to fight effectively in a number of tasks could achieve this balance. In defence, units should offer mutual support, create overlapping fields of fire and react aggressively to throw attackers off-balance. In the attack, Montgomery stressed the need to force the enemy to commit his reserves across a wide front, perhaps by probing; then to focus the attack on a narrow front, with reserves thrown into the breach.

15

Previous pages: The British victory at Trafalgar established a naval tradition for boldness – 'off-balancing' their enemies; however, it also resulted in the death of Nelson.

Off-balancing might take a number of forms, including disrupting or degrading enemy command and control, raiding (by organized armed forces rather than guerrilla groups), and covert operations or sabotage. Raiding is an off-balancing tactic with a long history: the Seljuk victory over the Byzantines at Manzikert in 1071 was preceded by off-balancing attacks by cavalry.

Raiding at sea has also been common. The defeat of the Spanish Armada (1588) was in part due to harassment by the English ships, as well as an off-balancing attack with burning vessels as the Spaniards lay at anchor. In the Second World War the British Commandos' St Nazaire Raid (1942) was designed to disrupt the ability of the German navy to use its U-boats. On land, the British LRDG (Long Range Desert Group) and the SOE (Special Operations Executive) and the American equivalent, the operational groups of the OSS (Office of Strategic Services), set out to cause as much disruption as possible to the Axis forces. Their achievements led to the growth of 'Special Forces' in a number of armies across the world. In the Falklands (1982) the British Special Air Service (SAS) demonstrated that off-balancing raids sowed uncertainty and fear among opponents while also pinning down significant manpower to defend key installations.

A classic example of the ability to constrain or disrupt an enemy can be found in the age of sail at Trafalgar. Admiral Nelson demonstrated that by throwing the enemy off-balance he could achieve a decisive victory against superior numbers.

Trafalgar, 1805

When hostilities resumed between Great Britain and France in 1803, Napoleon mustered an army of 590,000 men, many of whom he concentrated at Boulogne on the Channel to invade England. To cross safely, he would first have to neutralize the Royal Navy. By concluding treaties with continental powers to acquire ships from Spain, the Netherlands and Genoa he hoped to build a force equal to the maritime force of the British. In 1805, he had 64 vessels scattered between Toulon in the Mediterranean and Texel on the North Sea, but they were pinned in their ports and unable to concentrate.

The Royal Navy, keeping its vessels at sea through an efficient system of repair and refitting, stationed flotillas close to the mouths of enemy ports. The key strength of the British strategy was to maintain a strong force in the western mouth of the Channel off Ushant. If any French ships managed to break through a blockade, the British force there would immediately sail to Ushant to reinforce the fleet and prevent any French attempt to assemble. Thus the Channel and the shores of England remained secure.

Napoleon ordered Vice Admiral Villeneuve, commander of the French fleet at Toulon, to extricate and assemble the separated squadrons. To draw the Royal Navy away, Villeneuve was to sail for the West Indies and inflict as much damage as possible. Napoleon reasoned that, lacking vessels to protect the West Indies *and* maintain the blockade, the British would have to abandon the latter. By sailing quickly back to Europe, the French could then link with the other European fleets, numbering about 40 vessels, destroy the depleted British forces in the Channel, and permit Napoleon's invasion force to land in southeast England.

Rear Admiral Horatio Nelson, already a national hero for his glittering naval victories, was convinced that the focus of the French strategy was the Mediterranean. When Villeneuve sailed from Toulon with 11 ships of the line and 9 frigates in January 1805, Nelson thought he was making for Egypt, and his instinct was to intercept him. A second French sortie in March again persuaded Nelson to station himself around Malta. In fact, Villeneuve was trying to co-ordinate a break out to Martinique with the Franco-Spanish ships in Cadiz under Admiral Gravina. As soon as Nelson grasped the French strategy he set off in pursuit across the Atlantic, determined to attack. Villeneuve was so unsettled by Nelson's arrival in the West Indies that he disregarded fresh orders from Napoleon that he should remain in place for at least 35 days to permit the French fleet at Brest to escape the blockade and join him. The plan was for the combined fleets then to sail back and open the other blockaded ports before sailing for Boulogne.

Horatio Nelson was already an experienced and audacious fighter by the time of Trafalgar in 1805.

Villeneuve did not wait, however. He made for the Bay of Biscay, where an unsuccessful encounter off Cape Finisterre with a small British squadron reinforced his decision to ignore Napoleon and head for Cadiz.

Meanwhile, Nelson had returned to Gibraltar and planned for the inevitable confrontation with the combined Franco-Spanish fleet. In his Secret Memorandum he developed further a style of operations he had made his own. Nelson believed in closing with the enemy with urgency and aggression. The aim was to engage the enemy, to fight rather than manoeuvre, with his officers taking the initiative without waiting for signals to be flown. This was in stark contrast to previous expectations of naval warfare. Most commanders sought to control their captains, compel them to follow their flagship 'line ahead' and bring the maximum firepower to bear in a broadside. If possible, commanders might seek to cross the head of an enemy line. Villeneuve had to give strict directions to his inexperienced officers and crews, but Nelson was confident that his men had the skills and background for greater flexibility.

Nelson's Secret Memorandum was, in fact, well known. Villeneuve even predicted it: 'The British fleet will not be formed in line of battle parallel with the Combined Fleet ... Nelson ... will seek to break our line, envelop our rear, and overpower with groups of his ships as many of ours as he can isolate and cut off.'[2] Nelson believed that it would take too long to assemble 40 ships into line ahead formation, and that blazing away with broadsides would invite heavy casualties with no guarantee of success. Instead he proposed to form two columns to close on the enemy line at right-angles. Rear Admiral Collingwood, his immediate subordinate, would attack the last 12 ships, while Nelson would lead another group into the centre. This would force the head of the enemy column to bring their ships about, which would take time and allow Nelson to neutralize the enemy's numerical advantage.[3]

When Nelson described this audacious plan to his officers, 'it was like an electric shock. Some shed tears, all approved ... it was new – it was singular – it was simple.'[4] The risk was that, for the last hundred yards, the British ships would be exposed to a galling fire, but Nelson claimed: 'I think it will surprise and confound the Enemy'.[5] Nelson knew his fleet's gunnery was far better than that of the French: the crews had trained hard and could manage three broadsides in five minutes, double the French rate of fire; his captains had spent years perfecting the handling of their ships, while the men were battle-hardened and eager for action. By bearing down and then piercing the enemy line, Nelson predicted this would throw the enemy Combined Fleet off-balance, pin them to a fixed position, and 'bring forward a pell-mell battle', and, he added, '*that* is what I want'.[6]

In mid-October 1805, Villeneuve put to sea again, soon pursued by Nelson's ships. Small frigates caught sufficient glimpses of the Combined Fleet to keep Nelson informed as to their bearing. At dawn on 21 October, the French were formed in line ahead about 10 miles (16 km) distant. Nelson gave orders to form the two columns at 0610 hrs and they began to press on towards the Combined Fleet. The winds were light, and when Villeneuve sighted the British ships, he ordered his entire fleet to go about in the hope of gaining the refuge of Cadiz. The manoeuvre left the Spanish and French ships, which had been deliberately mixed together, in some disarray; it also affected morale since it seemed they were on the run. Gaps appeared in the line, while other ships were bunched together making it impossible to bring their guns to bear.

Nelson wanted to thwart this retreat and altered course to cut the enemy line higher up than originally intended. Collingwood also swung further to the southeast so that vessels in his column now sailed independently towards the French in a ragged line. At about 1145 hrs the French opened fire. Collingwood's ship, the *Royal Sovereign*, was first to pierce the line, taking a broadside from the *Fougueux* and *Santa Ana*. However, as *Royal Sovereign* passed under the stern of *Santa Ana* and ahead of *Fougueux*, she raked both with double shotted guns, causing significant damage. After 40 minutes of close engagement, *Royal Sovereign* had taken a great deal of punishment, but *Santa Ana* surrendered. One by one the other ships of Collingwood's column crashed through the enemy line, causing havoc. The Combined Fleet was unable to manoeuvre and was now at the mercy of the British guns.

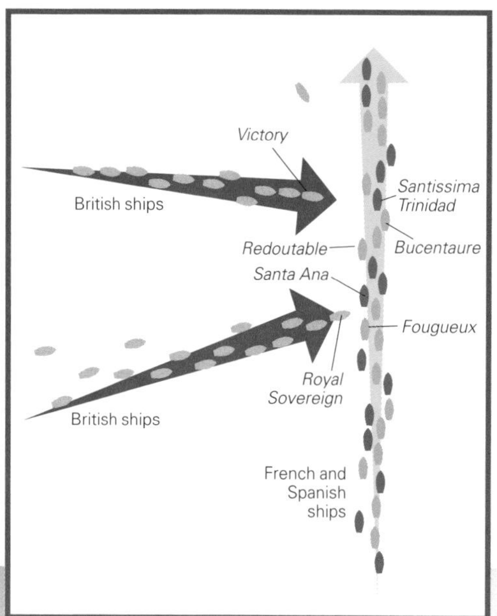

Nelson, aboard *Victory*, led the first three ships of his column towards the enemy line, and turned to starboard on making contact so as to bring as many guns as possible on his three-deckers to bear. The damage they inflicted was tremendous. *Victory* broke through the line heading for the *Santissima Trinidad*, the largest ship in the enemy fleet. Taking a broadside from the ships nearby, *Victory* fired into the stern of the *Bucentaure* with a 68-pounder mounted on the forecastle, loaded with round shot and a keg of 500 musket balls. Double shotted guns were also fired at close quarters into the cabin windows:

the iron and lead projectiles smashed through the decks and caused mayhem. Midshipman Badcock later wrote: '[the *Santissima Trinidad*] had between three and four hundred killed and wounded, her beams were covered in blood, brains and pieces of flesh, and the after part of her decks with wounded, some without legs and some without an arm.'[7] *Victory* now closed on *Redoubtable*, and their rigging became entangled, locking the ships together. Guns were fired at point blank range into the two hulls. Tragically, Nelson was mortally wounded by a sharpshooter as he strode the deck, but the balance had already tipped in his favour. The arrival of each British ship added its weight of fire, contributing to an overwhelming success.[8]

Opposite:
Nelson's plan at Trafalgar was to avoid a pounding of broadsides in a line-ahead formation: he broke all naval conventions by piercing the enemy line and destroyed them ship to ship.

Analysis

Nelson's plan was bold and daring, but was based on certainty about the capabilities of his men and his weapon systems. He appreciated that the Spaniards and French would fight hard and that their numerical superiority would be a significant problem, but he also knew that an aggressive and off-balancing attack would unsettle the enemy to the point of collapse.

Nelson's attack also depended on the inability of the Combined Fleet to respond effectively. By piercing the line in two places, and by isolating the rear, he pinned the French and Spanish in their formation. He had to maintain the pressure since a single blow would not be sufficient. The arrival of each British vessel ensured, as the military historian and strategist J. F. C. Fuller put it, a 'continuous concentration of fire on the enemy's ships' at the critical point.[9] Many commanders have understood the need to *continue* to off-balance the enemy, to keep him moving and keep on top of him. Nelson had pursued Villeneuve and unsettled him time and again, while creating confidence in the minds of his own men. The warm and enthusiastic reaction to his plan was probably not an exaggeration. Montgomery called this the 'atmosphere' a commander must inculcate – a combination of inspiration and firm guidance – but the audacity and pugnacity of Nelson's idea was its very appeal. Nelson initially off-balanced the Combined Fleet and maintained the pressure throughout the action, resulting in a decisive victory.

16

The Overland Campaign, 1864–65

Mass

Numerical superiority can prove decisive in battle: sheer weight of numbers may overwhelm opponents. At the tactical level, *local* superiority of numbers constitutes an effective concentration of force (see **13**), which is crucial for the ratios needed to overrun defended localities – usually estimated as a minimum of three to one. At the higher strategic level, superiority can also be manifested in industrial capacity.

Many armies in history augmented their core of professional or semi-professional forces with larger numbers of less well-trained and cheaper levies and auxiliaries. Tiglath-Pileser III, the Assyrian emperor of the mid-8th century BC, built up a substantial composite army to overrun Babylonia and Urartu. The Roman Republic survived the crisis of the Second Punic War, as Polybius noted, through superior manpower reserves alone since, despite Hannibal's crushing victories on the battlefield, new armies were quickly reformed. The Roman Empire's ability to deal with opponents on every frontier, with varying degrees of success, was due in part to its mixed forces, drawn substantially from locally recruited units, but also to its internal economic system, road network and political-military organization.

16

Previous pages: General Grant's supply base in 1864. The sheer volume of arms and stores indicates the logistical superiority of Union forces in the American Civil War, which contributed to their victory in the field.

In the early modern period, the Holy Roman Emperor Charles V could in theory field five distinct armies and fleets, and possessed the ability to deal with rebellions and wars on every frontier. The Imperial military machine in the Thirty Years War (1618–48) could put large numbers in the field, assisted by new fiscal and political organization in the early modern European states. In the 19th century, Napoleon's Grande Armée gave him the flexibility to defeat coalitions of continental enemies, and then garrison his conquests.

However, by the 20th century, the mechanization of war meant that superiority of mass was also manifested through industrial capacity, which became more important than the 'Big Battalions' of manpower: what the Germans termed the *Materielschlacht* (the battle of materiel) was vital.[1] On the Eastern Front of the Second World War (1941–45) the quantity and quality of, and the rapid ability to replace losses in, armour, mechanized transport and aircraft, turned the bloody engagements of Kursk (1943; see p.164), Poland (1944) and Berlin (1945) into Soviet victories. This was also true for Allied global and total war strategy in the Second World War.[2]

The Overland Campaign, 1864–65

In the American Civil War Lieutenant General Ulysses S. Grant had been so successful in the campaign in the Western Theater against the Confederate forces around Vicksburg in 1863 that he was appointed to command the entire Union army. On taking the field in spring 1864, he dispatched his two subordinates, William Tecumseh Sherman and George Meade, to pursue and engage the Confederates wherever they found them, while he led an army of 120,000 towards Richmond in Virginia, the Confederate capital. Opposing him, Robert E. Lee, the pre-eminent Confederate commander, could field just 60,000, as Grant was aware.

Grant intended to overwhelm the Confederates by sheer weight of numbers, knowing that he could sustain heavy losses if necessary. Confident that Sherman could brush aside the smaller Confederate forces he encountered, he had given instructions for him to 'get into the interior of the enemy's country as far as you can, inflicting all the damage you can against their war resources', while the abundance of Union troops and ships meant that further diversionary operations could be mounted at Mobile and in the Shenandoah Valley.[3]

Major General William T. Sherman (seated, centre) and his staff in the bloody campaign of 1864.

Grant encountered Lee's army in the Wilderness country, near the Rappahannock River, and for two days the contending forces struggled through the tangle of woods and brush until Grant's men were repulsed on 6 May. Lieutenant Colonel Horace Potter, a Union officer, recalled that: 'All circumstances seemed to combine to make the scene one of unutterable horror – it was as though Christian men had turned to fiends, and hell itself has usurped the place of earth.'[4] Smoke obscured the location of enemies. Small parties that chanced upon each other shot, bayoneted and clubbed each other to death. Parts of the woodland caught fire and burned the wounded. A federal officer wrote: 'Each man fought on his own resources, grimly and desperately.'[5] The Union troops had inflicted losses of 10,000 on the Confederates for a loss of 18,000 of their own.

Striking again to the southeast, from 8 to 19 May Grant's army fought a series of engagements, known as the Battle of Spotsylvania Court House. These were intended to turn the Confederate's flank and cut their retreat towards Richmond, but also to inflict as many casualties as possible. Once again Grant lost twice the number of men as his enemy. Undeterred, he thrust again to the east and more bitter close-quarter fighting followed in the thickets of the North Anna stream and Pamunkey River. Grant wrote determinedly to the government in Washington: 'I propose to fight it out on this line if it takes all summer'.[6] Lee, fighting at Cold Harbor in early June, admitted that he had no reserves left: 'Not a regiment – and that has been my condition ever since the fighting commenced. If I shorten my lines to provide a reserve he will turn me; if I weaken my lines to provide a reserve he will break them.'[7] However, Lee's men, depleted and on inadequate rations, but with an unshakeable spirit, still held the Union forces. Grant was again checked. With 7,000 casualties lying before the Confederates' shallow trenches, some of the Union soldiers had even refused to renew the attack.[8] Lee's soldiers inflicted losses on the Union equal to their own strength in just one month, yet the Union army was still intact and capable of engulfing what was left of the Confederacy.

Grant now moved his entire force southwards, crossed the James River, and threatened Richmond from the east. Lee's army could not prevent the manoeuvre, but the garrison at Petersburg, with its extensive field fortifications, managed to hold the Union army. When Grant tried to move further south, Lee checked him once more. However, it was becoming clear that the Confederacy could not be strong everywhere. The Union seemed to be able to produce limitless numbers of men, munitions and equipment. By contrast, in the south, everything was in short supply: rations were so scarce that green corn and roots were staple fare; when dyes for grey cloth began to run out, uniforms were stained with the local 'butternuts' instead. The situation was aggravated by Sherman's drive through the southern states, where he used his superior numbers to outflank his experienced adversary General Joseph Johnston, turning him out of successive strong positions.[9] Because of this, Johnston was replaced by John B. Hood, who made a series of sharp assaults to arrest the retreat. For all their ferocity, the Confederates suffered losses of 10,000 at Decatur and 5,000 at East Point – men it would be hard to replace. Hood was ordered to resume a defensive posture, but, by the end of this campaign, 35,000 Confederates had fallen. Sherman pressed on, took Atlanta and then marched through Georgia, laying waste to communications, crops and industry in order to cripple the South's economy.

The Confederates continued to mount successful counter-offensives. Hood penetrated Tennessee until checked at Nashville, while General Early pushed into the Shenandoah Valley until he was ejected by General Philip Sheridan.

Opposite: Grant maintained his relentless pressure of numbers throughout the entire Overland Campaign, supported by a vast logistical chain. Railroads were crucial to the supply of all armies in the Civil War – this railroad mortar is at Petersburg, July 1894.

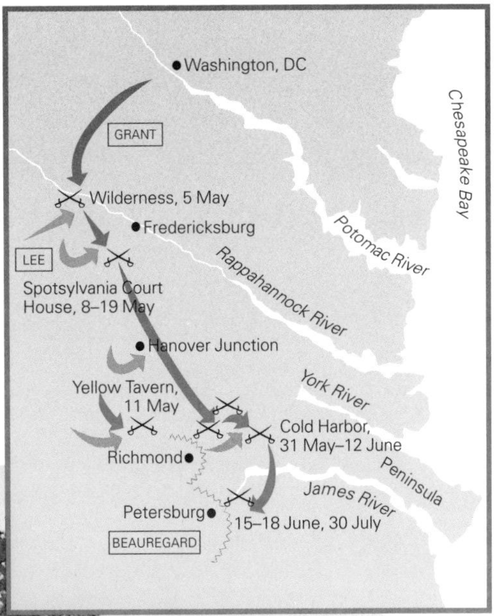

An attempt by Union forces to break through at Petersburg again was repulsed when their troops failed to secure the crater of a gigantic mine detonation.[10] Even in late 1864, Lee could still consider a plan to bring together the Western Confederate army (once again under Johnston) and his own battered force, to defeat Sherman and then confront Grant's overwhelming numbers. However, Nathan Bedford Forrest, another Confederate commander still in the field, understood the importance of mass, allegedly stating that victory depended on getting to the centre of gravity 'fastest with the mostest' – and the Union was completely superior in this regard. As historians Curt Johnson and Mark McLaughlin put it: 'Grant had one resource above all others which the South could not match: manpower. In the war of attrition he decided he must wage, the cruel arithmetic of numbers would eventually give him victory.'[11]

The thinning rebel lines at Petersburg were finally turned in the spring of 1865, and the Confederates were defeated at Five Forks in April.[12] Grant brought to bear the full strength of the Union army, and his greater numbers enveloped the 27,000-strong Army of Northern Virginia. Consequently Lee abandoned Richmond, to be pursued by three times his own number; as he tried to escape southwards, a Union cavalry corps under Sheridan, almost the size of his entire army, swept around him. What pitiable rations the Confederate soldiers had left were cut off when this Federal cavalry captured their baggage trains. Lee was aware that the Union armies were now on three sides and closing in. Further south, the situation was similar, as Sherman's

Richmond in ruins: Grant overwhelmed the Confederacy by weight of numbers and brought the rebellion to its knees in 1865.

army curled around the remaining, much-diminished army of Johnston. Lee wrote to the Confederate President Jefferson Davis: 'I cannot see how we can escape the natural military consequences of the enemy's numerical superiority'.[13] Bowing to the inevitable, Lee capitulated at Appomattox, accepting the generous terms offered by Grant. Grant's use of vastly superior numbers had overcome even the most skilful of generals, illustrating the importance of mass in war.

Analysis

Overwhelming numbers sometimes produce victory, but are not necessarily sufficient by themselves, since any force requires the materiel of war, intelligent leadership, good motivation and preparatory training. There is always the risk that numbers comprise less well-trained and inexperienced troops, who will suffer heavily. High losses meant that Grant's army had begun to reach its limit in the Wilderness campaign, and his return to manoeuvre helped win the final victory. But the numbers had done their job in inflicting unsustainable losses on the Confederates, thereby permitting a vast envelopment that even an exceptional commander like Lee could not prevent. When combined with superior firepower, determined leadership and men with high morale, an army of mass can prove a decisive force.

17

Alesia, 52 BC
Kursk, 1943

Defence in Depth

Defence in depth is a tactic applicable to the theatre of a campaign as well as to individual battles. It is a concept which is invoked p in strategic discussions of the defence of the late Roman Empire: the balance of troop dispositions changed from an emphasis on preclusive deployment along the frontiers to a more mixed arrangement consisting of frontier units supported by garrisons in the interior at significant cities and important communication points, with some central forces held in reserve for allocation as appropriate. Nevertheless, in open battle the importance of morale and confidence are such that a force which enters an engagement in a defensive mentality immediately places itself at a disadvantage, so that success in the field solely through resilient defence is a rarity.

Previous pages: Soviet forces, using the cover of smoke and protected by armour, defended Russia in depth and defeated the German offensive at Kursk during the bitter campaign of 1941–45.

One exception is the sea battle off the Arginusae Islands in 406 BC during the war between Athens and the Peloponnesian League of allies led by Sparta (431–404 BC). The Athenians, after a defeat at Notium (near Ephesus in modern Turkey) and the blockade of the remnants of their Aegean fleet in the harbour of Mitylene by superior Peloponnesian forces, had to prevent the Spartans from seizing complete control of the eastern Aegean and thereby interrupting Athenian grain supplies through the Hellespont. At Athens 60 triremes were launched, possibly brand-new hulls built with wood recently imported from Macedonia and rowed by scratch crews including both slaves and the richest citizens, for whom such manual labour was abnormal. These joined 90 ships at Samos and proceeded north towards Lesbos to relieve the blockade at Mitylene. The Spartan admiral Callicratidas withdrew 140 of the Peloponnesian ships from Mitylene to confront this threat, meeting the Athenians near the Arginusae Islands. The Athenian disposition, in two lines, was unusually defensive, either because of recent defeats or because the inexperience of many crews and the newness of the ships from Athens meant they had not yet achieved prime fighting condition. In the ensuing battle they hoped to prevent any Peloponnesian ships that broke through the front line from turning sharply to perform the classic *coup de grâce* of disabling the steering gear and oars. The Athenian tactics worked, although losses were high on both sides, with 77 Peloponnesian and 25 Athenian ships destroyed.

Another context in which defence in depth plays a part is the siege, especially as defensive constructions became more complicated in response to the greater capacity of assailants. In a siege the successful repulse of an attack can boost defenders' morale while weakening that of the besiegers, and a succession of barriers that have to be overcome by the attackers may increase the chances of the defenders being able to inflict a reverse. Defence in depth can also be achieved when supplementary internal walls are rapidly built to replace sections of a main perimeter that has been destroyed or compromised. At Perinthus in Thrace in 340 BC, as Philip of Macedon's powerful assault destroyed more and more of the fortifications, the citizens desperately blocked up the lower storeys of adjacent houses and barricaded the streets to present so many extra lines of resistance that Philip eventually desisted.

The most powerful cities came to be equipped with multiple lines of defence. Thus Constantinople ultimately had four: the Long Walls ran from the Black Sea to the Sea of Marmara about 50 miles (80 km) from the city, while the main defences consisted of a moat lined by a breastwork, a substantial outer wall, and the main inner wall 32 ft (10 m) high with towers of 65 ft (20 m) located every 164–196 ft (50–60 m). These defences saw off the threats from the Avars in 626 and the Arabs in 717–18 and were not breached until the Ottoman attack in 1453.

Alesia, 52 BC

Julius Caesar's investment of the Gallic hill-fort of Alesia (30 miles/51 km, northwest of Dijon in central France) in 52 BC provides a good example of a siege which included a pitched battle and defence in depth. As the Romans were consolidating their authority across Gaul, the Gallic tribes, led by the Arvernian Vercingetorix, had risen in a final bid for liberty. A series of Roman successes led Vercingetorix to take refuge at Alesia, which Caesar then proceeded to invest. The Roman troops had to be distributed in seven camps along a perimeter of about 11 miles (18 km) around the fort, but it was clear that this was an insufficiently tight blockade to prevent the Gauls from communicating with their fellow tribesmen outside and receiving supplies. Caesar therefore began to link the camps with a connecting rampart 6 ft (1.8 m) high, reinforced by a perpendicular-sided trench, 23 ft (7 m) wide, and frequent towers, with 23 forts to provide added protection. He was also aware that Vercingetorix had summoned help from the Gallic tribes, and so along a perimeter of 14 miles (22.5 km) an outer ring of comparable defences, known as lines of circumvallation, had to be built facing outwards.

Caesar did not have sufficient troops to man these massive defences in strength, and so he set about making it harder for the enemy to approach the rampart, especially from the inner, Alesia, side. To prevent the Gauls from hurling javelins at the main wall, he had two further trenches, each 16 ft (5 m) wide dug. Where possible, the inner trench was filled with water and was further protected by the construction of another rampart topped by a battlemented breastwork. Frequent sorties demonstrated the need to step up

Reconstruction of a section of Roman moated palisade and towers at Alesia.

these protective measures, and so Caesar commissioned further obstacles. Shallower trenches, 5 ft (1.5 m) deep, were dug and armed with sharpened tree trunks or large branches, tightly packed five deep and with interlaced points, so that any attackers risked impalement as they struggled across. In some sectors diagonal lines of pits were dug, each equipped with a stout, sharpened stake rammed into position; these were arranged in groups of eight rows. In addition, short blocks of wood studded with iron hooks were sunk into the ground.

As food within Alesia began to run short, Vercingetorix sent the non-combatants out of the hill-fort in order to make the supplies last longer. Their pleas to be accepted as slaves and let through were refused by the Romans; they were denied re-entry into their home town. The defenders also debated the possibility of cannibalism. At last, a massive Gallic relief army approached, allegedly 250,000 strong and with 8,000 cavalry. Caesar's preparations would be tested to the limit. The first engagement was a cavalry battle in which initial Gallic successes were overturned by a charge of Caesar's German horsemen. The Gauls then spent a day preparing ladders and grappling hooks to overthrow the outer ramparts, and bundles of wood to fill in the ditches, before attacking one sector of the Roman line at midnight. The weight of their missiles had some impact on the Romans, but as the Gauls approached the main rampart they became entangled in Caesar's extra obstacles and the attack lost momentum. Meanwhile, Vercingetorix had attempted to attack the same sector from the inside, but had managed to get no further than filling in the first stretches of trench before day began to break. At that point, the Gauls on the outside called off their assault for fear of being attacked in the flank by Roman support units.

Thwarted twice by the lines of Roman defences as well as determined resistance, the Gauls then paused to reflect on how best to dislodge the blockade. A weak point at the northern end of the circumvallation was identified where a hill had not been incorporated fully within the Roman line. The Roman camp there was located on a slope, making it possible to assail it from above. Some 60,000 Gauls were said to have been allotted to the attack on this hill, which they began at midday; Vercingetorix, having observed their moves, launched an all-out sally to coincide. The Gauls approached the Roman outer defences with locked shields, while the various obstacles were gradually filled in with earth; weight of numbers began to tell and Caesar dispatched 3,000 men to sustain the defence under his most trusted commander, Labienus. These were not enough to save the camp, however, which was gradually abandoned as the Gauls started to tear down the turrets and ramparts. Meanwhile, the Gallic sally from inside the fortifications had failed to make progress against the sections of the rampart in the plain, as these were strongly defended by the extra obstacles, but launched themselves on one of the steeper sections with some success; Caesar had to counter this with the redeployment of troops from quieter sectors.

The crux of the battle was still the northern camp and Caesar was eventually able to proceed there himself, accompanied by 2,000 men withdrawn from elsewhere on the circuit. He also instructed his cavalry to ride round behind the Gallic attackers, so that they were threatened from both sides. The approach of Caesar, easily recognizeable in his scarlet cloak, galvanized both sides, as they appreciated that this was the critical moment for the whole blockade, and hence the fate of Gaul. The contest was fierce, but was decided when the Gauls saw the Roman cavalry behind them and turned to flight. Gallic losses were substantial, and Vercingetorix, realizing that there was now no chance of relief or escape, surrendered Alesia.

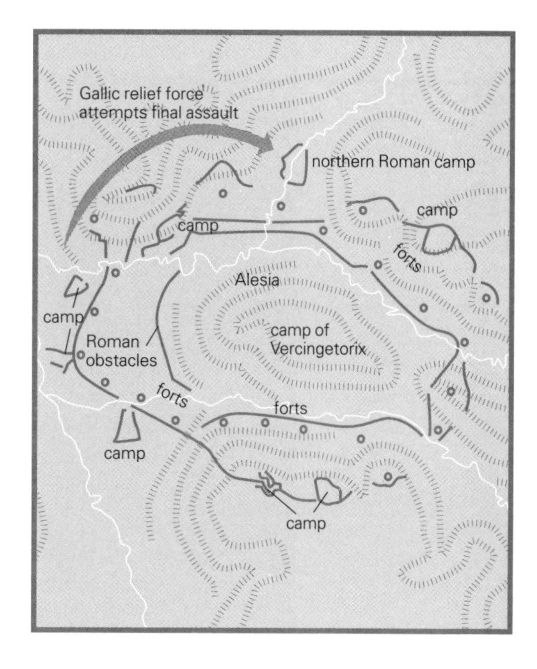

Alesia, showing the hill-top fort with Roman defences and circumvallation.

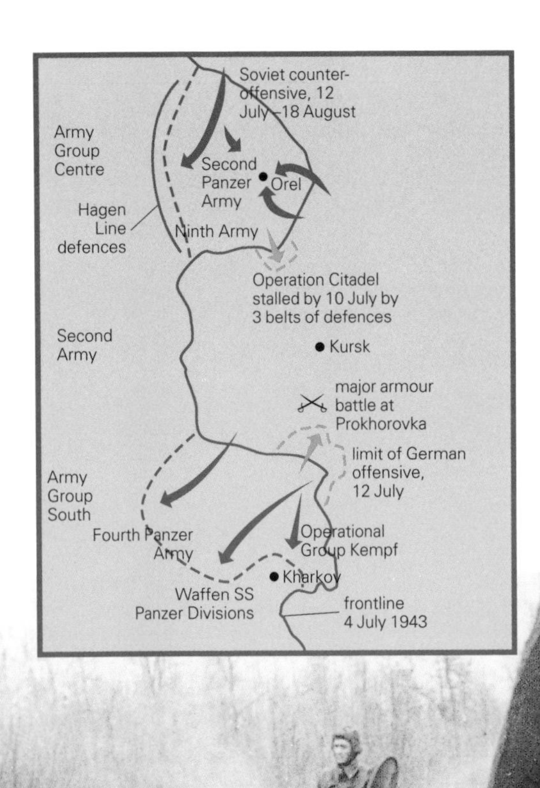

Kursk, 1943

The concept of defence in depth has been used on many subsequent occasions. The key point is that it enables a force to absorb the shock of an enemy attack, and dilute the concentration of its firepower. At the Battle of Kursk in 1943, the Soviets arranged their forces in a series of defensive zones each 3 to 4 miles (5–6.5 km) deep, up to a total of 25 miles (40 km), bristling with anti-tank weapons, minefields and artillery.[1] In the north of the battlefield, small groups of Soviet soldiers, armed with anti-tank weapons, occupied slit trenches inside the minefields with orders to disable as many German tanks as possible.[2] The Soviets chose not to hold the Germans on a particular line, but, anticipating large-scale German attacks, expected to give ground and permitted German armoured columns to drive into the defended zones. There counter-measures could be employed, including ground strafing by massed aircraft, to diminish the size and momentum of the push. Around these defensive strong-points they planned armoured counter-strokes. By these means, the Soviets blunted German attempts to cut off Soviet forces in a great salient, and turned the tables using massive reserves. German units found they were unable to offer mutual support between formations. By the end of the struggle, which was the greatest tank battle in history, the Germans were forced on to the defensive and driven back many miles.

Opposite:
At Kursk, the greatest tank battle in history, the Soviets absorbed the energy of the German thrusts, then drove them back beyond their start-lines.

Analysis

Caesar's account of the defence of Alesia in the *Gallic Wars*, the source of almost all our information, undoubtedly highlights the drama while underlining the efficacy of his preparations and actions. Archaeological work at Alesia from the mid-19th century onwards has provided corroboration for most of his descriptions of the defences and obstacles, although this has also revealed that Roman fortifications were not as complete as Caesar's narrative suggests and that the extra obstacles were not located along all sectors of the blockading rampart, since some parts were adequately reinforced by natural defences. Still, the siege represented a massive investment of effort. The intelligent elaboration of the Roman defences, combined with a clear strategy of holding the defensive line, enabled Caesar's army, which, though very large by Roman standards, probably numbered no more than 50,000, to preserve the blockade and so defeat the Gallic revolt. At Kursk, the careful siting of successive belts of defences, minefields and reserves meant that the impact of the German armoured thrusts could be absorbed. Here the defence in depth was so effective that the Soviets were able to halt the German attack entirely and then turn to their own counter-offensive.

18

Panipat, 1526
Yom Kippur, 1973

Strategic Offence and Tactical Defence

In the course of a battle, it is vital for any commander to read the tactical situation and the configuration of the ground, and make assessments all the time. For example, in an 'advance to contact' the attacking force may begin by moving 'tactically', that is, making use of the terrain to conceal and protect the troops while maintaining extreme vigilance for any signs of the enemy. Troops will move in 'bounds', short distances that can be supported by the fire of neighbouring or specialist units such as artillery or aircraft. The movement of each formation will be co-ordinated so that there is always 'one foot on the ground' – a unit deployed ready to offer covering fire – with these firebases being continually recreated as the troops advance. If the enemy offers resistance, the attacking troops will immediately return fire but also seek protection in the folds of the ground, in the use of smoke or in the application of supporting gunfire.

Commanders have to work quickly to locate the enemy, then focus their assets to pour the concentrated fire of their own, neighbouring or supporting units on to the enemy. In the course of an action, sub-units will often be compelled to switch from defence to attack, finding themselves moving with the utmost caution, concealed and camouflaged, then stationary and delivering fire on to enemy positions, and then, moments later, moving at speed to close with the enemy in a close-quarter battle. The aim is to 'win the firefight', a process that can last for hours. At Goose Green in the Falkland Islands (1982), for example, the British Paras took eight hours to suppress their Argentinian opponents before beginning their next phase: the assault.

18

Previous pages: An Israeli bombardment on the Golan Heights in 1973 as they make desperate attempts to stem the Syrian advance during the Yom Kippur War.

On campaign, the requirement to switch from one posture to another is the same. Indeed, the ideal situation is to move into a location that is strategically valuable, but, once there, to defend a strong position and force the enemy to deliver an attack. At the Battle of Panipat in 1526, and again in the Yom Kippur War of 1973, the effectiveness of this technique was exemplified.

Panipat, 1526

Babur, the 'Tiger', was a Chaghatai Turk who could claim descent from both Genghis Khan and Timur. Born in the Ferghana Valley of Central Asia, he spent many years from a very young age fighting to gain and maintain control over Samarkand and Kabul. From his base in Afghanistan he made several raids into the Indian sub-continent against the Muslim Lodi dynasty of Delhi. Their considerable military strength initially prevented Babur from launching a full-scale invasion, but in 1525 he resolved to defeat his rival, Sultan Ibrahim.[1] Babur could muster just 12,000 men against the Lodis' 100,000, but he had been quick to acquire the latest gunpowder weapons, which the Lodis did not possess. Nevertheless, the relatively slow rate of fire of his cannons and handguns, and the Lodis' superiority in cavalry, meant that Babur had to adopt a defensive posture and slow his opponents to maximize the effect of his weapons.

Babur advanced rapidly to Panipat, which lay before Delhi, knowing this sudden threat to his opponents' capital would prevent them seeking refuge behind their walls. He selected the battlefield carefully: in a relatively narrow space in a forested area close to the town he arranged a line of 700 wagons secured together by ropes of hide, with breastworks covering the small gaps between them. Behind this makeshift defence he packed his gunners and held back his cavalry as a mobile reserve. On the forest flank, he dug a ditch and filled it with logs to create an obstacle to cavalry. He knew that, eventually, the Lodis would have to advance. For a week there was a standoff. Babur's force was augmented by deserters, while the logistical problem of feeding large numbers was obviously greater for his enemy. Babur waited patiently behind his barricades.

Babur pushed into the Lodi Sultanate to induce them to attack, and fought a brilliant defensive battle at Panipat.

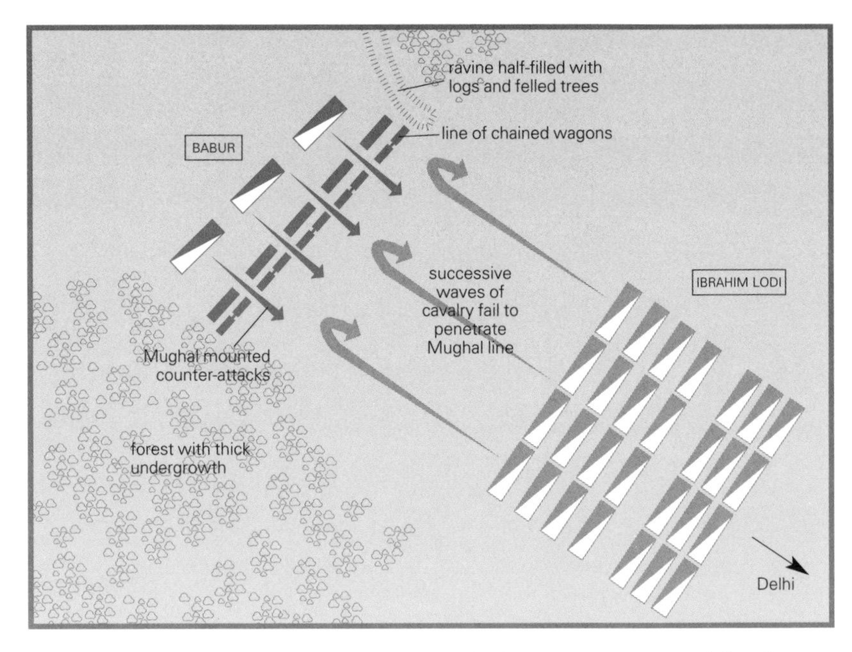

At Panipat, Babur channelled his impetuous enemy into hopeless charges, marshalling his own forces until the critical moment.

Babur believed that Ibrahim 'was unproved [that is, inexperienced] but brave; he provided nothing for his military operations, he perfected nothing, nor [knew how to] stand, nor move, nor fight.'[2] He anticipated a frontal assault, with the Lodis launching the full weight of their forces in a series of attacks.

The Lodis at last decided to advance. The army was preceded by numerous war elephants, perhaps as many as 100, and behind them came a mixed force of cavalry and infantry, most armed with a mace. Ibrahim Lodi believed he could sweep away Babur's defences and overwhelm the Mughal band through sheer weight of numbers. However, the Lodis were hemmed in by the settlement and forest, and when Babur's cannons were fired, the terrified elephants turned and began to stampede back through the densely packed troops behind. Smoke added to the confusion, and an attempt by the Lodi cavalry to attack the Mughals failed: they could not penetrate the wagon barrier. As they vainly tried to rein in, the Mughal gunners poured fire into them at close quarters. The accumulation of bodies and wounded horses added further to the confusion. At the critical point, Babur ordered gaps to be made in the wagon defences through which his cavalry rode, pursuing the Lodis and throwing them into disorder. As the Lodis fled, Babur sent a column of mounted men to Agra, where the families of the Lodi rulers had sought refuge. At a stroke, Babur had smashed his enemy's army, seized their capital and taken captive the ruling élite.[3] He now established a dynasty that would last almost 300 years.

Yom Kippur, 1973

After their humiliation in 1967 (see p. 37), Egypt and Syria sought revenge and wanted to regain the territory lost to Israel, namely the Sinai and the Golan Heights. President Sadat of Egypt believed a limited war against Israel would force them to the negotiating table. Hafiz al-Assad, Syria's head of state, was more belligerent, recognizing that the Golan Heights would have to be seized by force. Both states appreciated the need for a surprise attack and had been accumulating Soviet weapons and aircraft since the 1967 conflict. Egypt also embarked on a diplomatic offensive to ensure widespread international support in the event of a 'crisis'. Sadat himself hoped to win popular favour by fighting Israel. The Egyptian army were already engaged in a so-called 'war of attrition' against Israeli fortifications along the Suez Canal, but political support at home for Sadat was limited and relations with the Soviet Union were strained.

Preparations for the Egyptian attack were well concealed, despite an overt propaganda campaign. Israeli intelligence indicated that the Egyptians were not ready for war, since they were awaiting the arrival of more Soviet jet fighters and Scud missiles, and signs of military activity were dismissed. Even a direct warning from King Hussein of Jordan to the Israeli Prime Minister, Golda Meir, was disregarded, and a partial mobilization of reservists was only ordered shortly before the Arab attack.[4]

The Egyptians launched their attack on 6 October during the Yom Kippur holiday in order to catch the Israelis off guard. Forward units were screened with a large number of SAM (Surface-to-Air) missile batteries, which could neutralize Israeli close air support – these Soviet-built missiles destroyed

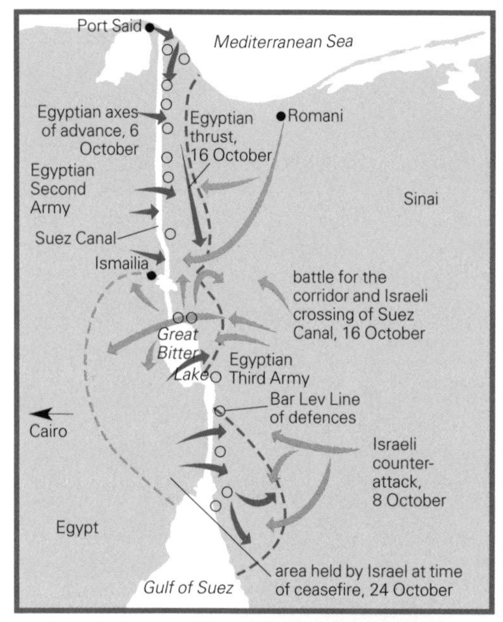

49 Israeli aircraft. The small Israeli garrisons on the Bar Lev Line of fortifications along the Suez Canal were enveloped, and giant water cannon were used to breach the canal's sand banks.[5] In a matter of hours the Egyptians were across the canal and into the Sinai. The Israelis counter-attacked with armoured formations thrusting from within the desert. Anticipating this, Egyptian infantry units were equipped with large numbers of anti-tank RPGs (Rocket-Propelled Grenades) and Sagger missile systems, which inflicted significant losses: the Israelis lost 500 tanks during the campaign. Only Ariel Sharon's 143 Armoured

Division enjoyed any success in preventing an Egyptian drive eastwards. The Egyptians then halted, reluctant to advance against Israeli positions in depth, beyond the limit of their air cover and the range of the SAM batteries.

On the Golan Heights, the Syrians launched their attack at the same time as the Egyptians. They had a substantial numerical advantage: five Syrian divisions faced just two Israeli brigades, with 1,300 tanks against 180. Syrian commandos first seized a key communications node at Jabal al Shaikh, and Syrian armour fought its way to the approaches of Nafah, an important junction astride the Golan Heights. Their carefully co-ordinated plan ensured that forward units did not stray beyond their air umbrella and SAM systems, but this rigidity meant that an early opportunity to capture Nafah was lost. At this stage, four days into the fighting, Israeli units could be deployed as more cohesive formations. The defence became more organized and resistance to the Syrians was strengthened.[6]

Both the Egyptians and Syrians had advanced, exploiting surprise and using an offensive strategy combined with the defensive tactics of SAMs and anti-tank weapons to neutralize the overall Israeli advantages in armour and air power. But they had reached the culminating point of their offensive, and here things started to go wrong. Sadat wanted to assist the Syrians, whose offensive had stalled. Despite opposition from his Chief of Staff, the Egyptian armour drove head on into the waiting, and now well-prepared, Israeli divisions in the Sinai. On 14 October, as many as 250 Egyptian tanks were knocked out. The attack lost momentum and units withdrew. The Israelis selected this moment to strike back, themselves now using the concept of strategic offence and tactical defence to defeat the Egyptians.

Recognizing the effectiveness of the Egyptian anti-tank systems, the Israeli offensive was spearheaded by infantry who infiltrated between Egyptian positions to attack from the flanks and rear. Sharon's armoured division struck at the junction between the Egyptian Second and Third armies near the head of the Great Bitter Lake. After fierce fighting, the Israelis broke through and raced for the Suez Canal. A small bridgehead across the canal

Opposite:
In the Sinai, after initial setbacks, the Israelis used a strategic offence and a tactical defence across the Suez Canal to halt Egyptian forces and thus steal a victory in 1973.

was established by infantrymen without armoured support. As the Egyptian armoured counter-attack went in, it was decimated by Israeli M72 anti-tank rockets. Meanwhile, infantry units continued to press forward to destroy SAM batteries. With the Egyptian armour temporarily checked and its air screen punctured, the Israeli air force and armoured formations could once again lead the offensive. After some difficulty, four pontoon bridges spanned the canal and the Israeli 162nd Division crossed and drove southwestwards. This strategic thrust cut off the Egyptian Third Army in the Sinai and practically opened the road to Cairo.[7]

In the Golan, there was debate about whether Israel should remain on the defensive once it had contained the initial Syrian attack. The Syrian border was well protected with minefields, anti-tank obstacles and strong points. Yet the imperative of the campaign was political: Israel had not only to defend and recover its territory, it also had to acquire advantages to take to the negotiating table. The Israelis therefore attacked across the frontier, managing in just three days to bite and hold a pocket of Syrian territory, the Bashan Salient. This put them within artillery range of Damascus – a position of strategic strength where they could remain on the defensive tactically. Despite Syrian and Iraqi counter-attacks , supported by token forces from other Muslim states, the Israelis could not be dislodged. By the end of the campaign, Israeli paratroopers and commandos had mopped up all the Syrian pockets on the Golan Heights. As the Syrians prepared a larger counter-offensive, and the Soviets threatened to intervene to protect Egypt, the United Nations brokered a ceasefire to end the war.

Analysis

The concept of a strategic offence combined with a tactical defence permits an army to manoeuvre into a position of advantage which it can then defend. The success of this approach can be found in many historical eras, for example Žižka's victory at Kutna Hora in 1421 or the German posture on the Western Front in the First World War, where the Allies were forced into years of costly attacks. At the tactical level, the concept also implies that units are 'active' in defence. Raids and sorties can recover the initiative for a defending force, active patrolling can gather intelligence, and both can demoralize an attacker.

At the Battle of Panipat in 1526, Babur's strategic offensive created a situation in which the Lodis had to attack, but his defensive position then maximized his tactical advantages and negated his weakness in numbers. In 1973, the Arab combination of strategic offensive with tactical defence from anti-aircraft and anti-tank batteries was sound. The political decision by the Egyptians to advance beyond the range of their anti-aircraft cover allowed the Israelis to neutralize this advantage. The Israelis then attacked strategically and crossed the Suez Canal before adopting a defensive posture from which they inflicted heavy losses. The same applied on the Syrian border, where the Israelis secured a strategic position in which they could remain on the defensive and cut down counter-attacks. Even though the superpowers were soon involved, the Israelis had already forced the Arab states into a position where they had to negotiate. The initial Arab offensive had, nevertheless, undermined the notion of Israeli invincibility engendered by the Six Day War (1967). Had the Egyptians been able to assume a tactical defence under the umbrella of their air defences, they might have achieved even more.

19

Hattin, 1187
Napoleon in Russia, 1812

Drawing the Enemy

One of the most important aims of every commander is to make his enemy fight on ground of his own choosing so that he can bring his weapons to bear with the greatest chance of success. This can be achieved by manoeuvre, by the combination of strategic offence and tactical defence (see **18**), or by luring an enemy into a predetermined 'killing sack', for example by a feigned withdrawal. At the Battle of Hastings (1066) the Normans made a feint attack and then withdrew precipitately down the slopes occupied by King Harold's Saxons, giving the impression of a rout. Encouraged by this apparent collapse, one wing of the Saxon army pursued. The Norman cavalry reined in, and, with the reserves, they enveloped the now disordered Saxons and crushed them. Charles 'The Hammer' Martel used a similar ruse of 'withdrawal' in his campaign against the Frisians at Amblève (716). He surprised the far larger force of Frisians by raiding unexpectedly at midday, when armies traditionally rested; the Frisians thought this was just a limited demonstration and chased after the raiders with enthusiasm, only to be counter-attacked and destroyed by Charles.[1]

19

Previous pages: Saladin's forces at Hattin waited patiently until the crusaders had exhausted themselves before striking the decisive blow.

Hattin, 1187

In 1187, Saladin, Sultan of Egypt and Syria, faced a difficult situation. He had taken on the mantle of champion of Islam against the Christian Franks who ruled Jerusalem, Tripoli and Antioch, but was making little real progress and had spent more time fighting Muslim rivals than these enemies of the Faith. Moreover, he had been thrown off-balance because in 1181 Reynald of Châtillon, lord of Kerak, had precipitated war between Saladin and the Kingdom of Jerusalem by attacking caravans crossing his domains from Egypt to Syria. In the ensuing conflict Saladin gained little, but was humiliated when Reynald mounted a naval raid in 1182 which threatened Islam's holiest places, Mecca and Medina. By 1187 Saladin was looking to breach the truce he had made with Jerusalem in 1183, and another caravan raid by Reynald provided the opportunity. He was also aware that many nobles in the Kingdom were deeply opposed to Guy of Lusignan, who had seized the kingship by a coup in 1186. Raymond III of Tripoli, lord of Tiberias, was chief among these and had actually concluded an agreement allowing Saladin's troops to cross his lands in return for support against Guy.

Saladin proclaimed *jihad* and raised an army of about 30,000, mostly horsemen. Its cutting edge consisted of Turkish and Kurdish *ghulams* – highly mobile mounted archers who could harass the enemy with their bows, yet who were sufficiently armoured to be able to close with the formidable western knights. On 1 May 1187 Saladin's raiders surprised and annihilated a Frankish force of 400 cavalry, led by 130 Templar and Hospitaller knights, at the Springs of Cresson. However, this victory resulted in a reconciliation between Raymond and Guy, and did nothing to solve Saladin's strategic problem.

The Kingdom of Jerusalem, while small, was firmly anchored by a number of well-fortified cities, particularly Acre and Jerusalem itself. To besiege any of these would be to invite counter-attack by the Franks' powerful field army, which could find support and shelter in the numerous strong castles they had built across the kingdom. In the previous operations of 1183 Saladin had invaded Galilee to draw the Franks into battle, but Guy, at that time Regent, had resorted to 'Fabian tactics', that is, shadowing the Muslim army until it was eventually forced to disband. Saladin noted that Guy's military leadership on this occasion was very hesitant. He now reasoned therefore that if he attacked Tiberias, a rather isolated city on the Sea of Galilee, its loss would be a major blow to Guy and make him even more circumspect.

The Frankish army led by Guy was assembling at Saffuriyah, 16 miles (26 km) west of Tiberias, on the northern edge of the rich valley (Sahel al-Battof) that carries the Acre–Tiberias road. Their gathering was rather slow, as it took time for an army of 18,000 men (including 1,200 knights) to come together. Of course, this concentration of manpower denuded the cities and castles of the kingdom of their garrisons. Tiberias fell to the Muslims on 2 July, though its citadel held out, precipitating a debate among the Frankish leaders.[2] The result was that on 3 July the Franks marched out from Saffuriyah to face Saladin. We do not know precisely what Guy intended, but his army was formed for a 'fighting march'. The cavalry were divided into three divisions, with Raymond commanding the vanguard, Balian of Ibelin the rearguard, and Guy in the middle. These divisions were surrounded by a strong cordon of infantry and archers, which would enable the Franks to fight off the harassing fire of Turkish horse archers and to choose the moment to launch their decisive charge when the enemy presented a solid target.

An animated statue of Saladin in Damascus; at Hattin he drew Guy of Lusignan into open territory, which suited his highly mobile tactics.

But the Frankish decision to march out was deeply flawed in two key respects. First, Guy seems to have been unaware of the sheer size of Saladin's army, and, second, the Franks had embarked on a march through a waterless land. Initially Guy moved 7.5 miles (12 km) northeast to the springs at Tur'an on the northern edge of the Sahel al-Battof, a strong position, comparable to Saffuriyah but closer to the enemy. He then turned east against Saladin's army, perhaps seeking to draw it into the valley; here the Franks might be able to pin Saladin's force against the north–south ridge where it mounts to the higher ground to the east near Maskana. If this failed, the crusaders could retreat to Tur'an, repeating the manoeuvre at will and checking Saladin.

Saladin saw the decision to leave the springs of Tur'an as the turning point, later writing: 'But the devil seduced him into doing the opposite of what he had in mind and made to seem good to him what was not his real wish and intention. So he left the water and set out towards Tiberias.'[3] Saladin sent very strong forces to surround the Franks and to occupy Tur'an, completely cutting them off from water. He harassed the marching column constantly and sent more units against the rearguard, while keeping his main force sufficiently distant to prevent the Franks unleashing their charge.

That day the Franks, cut off from water, struggled in the July heat and made only some 2–3 miles (4–5 km) eastwards up the slope to Maskana where there was a little water to slake their thirst. Their rearguard, under severe pressure, became disorganized. As their difficulties multiplied, the leaders discussed whether to press on and in what direction. Eventually they spent the night in a dry and hostile land, surrounded by the Muslims, who set fires to aggravate their thirst. In the morning of 4 July both sides deployed for battle, but the Muslims held back and did not resume their attacks until the heat of the day

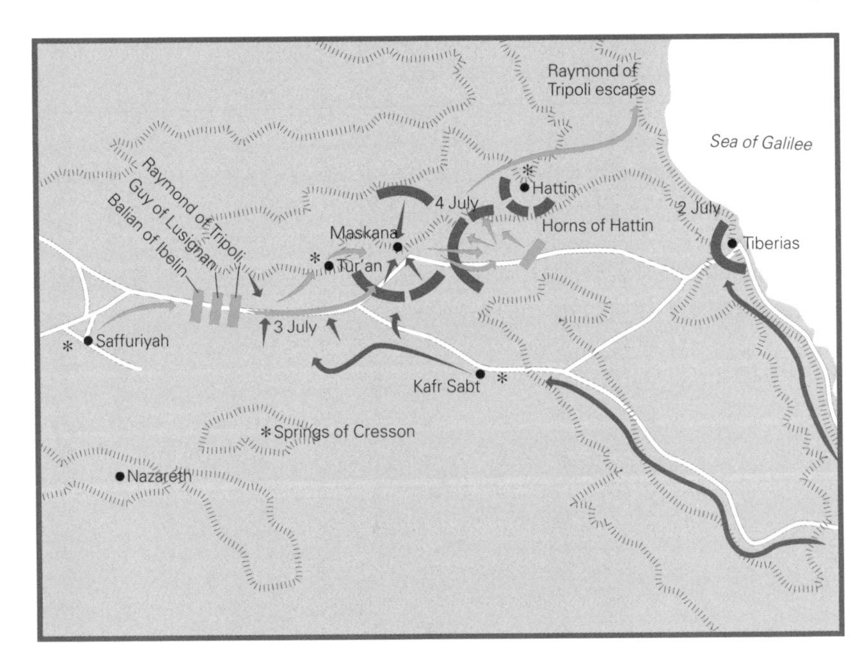

At Hattin Saladin was able to deprive the Franks of vital water sources and drew them into a long march on to ground of his own choosing.

and lack of water were starting to enervate their enemies. Quite possibly they also feared that a sudden break-out by the Franks might pierce their ranks.

What happened that day is not at all clear. At first the Franks held together; the Templars then tried unsuccessfully to break out and Raymond of Tripoli led a charge which the Muslims let through, allowing him and his limited forces to escape. The key event seems to have been the mass desertion of the demoralized infantry who fled to an extinct volcano, called, because of its shape, 'The Horns of Hattin'. Guy and the remnants of his army established themselves on its lower slopes and launched two savage charges, which almost reached Saladin, as his son recalled:

> When the King reached the hill with that company, they launched a savage charge against the Muslims opposite them, forcing them to retreat to my father. I looked at him and saw that he had turned ashen pale ... Then the Muslims returned to the attack against the Franks and they went back up the hill. When I saw them retreating with the Muslims in pursuit, I cried out in joy 'we have beaten them'. But the Franks charged again as they had done before and drove the Muslims up to my father. He did what he had done before and the Muslims turned back against them and forced them up to the hill. I cried out again 'we have beaten them'. My father turned to me and said: 'Be silent. We shall not defeat them until that tent [Guy's] falls.' As he spoke, the tent fell.[4]

After this failure, Guy tried to consolidate a defensive perimeter. He ordered tents to be pitched to create a fighting camp, but the survivors were gradually overwhelmed.

We have no clear picture of Guy's intentions in coming out to fight. But what is evident is that Saladin had drawn the Franks on to ground of his own choosing, where the speed, manoeuvrability and numbers of his army could be made to tell against enemies deprived of water. Almost the whole Frankish army was killed or captured, including Guy, while Reynald of Châtillon was executed by Saladin himself. Guy and the noble survivors were ransomed, and the poorer soldiers were enslaved, causing the price of slaves in Damascus to fall to 3 dinars. The fighting monks of the Orders of the Temple and Hospital, some 300 in all, were executed. But the real consequence was that, with the Frankish field army destroyed, the cities and castles of the kingdom could be picked off easily. On 9 July Acre fell, and while Tyre and a few other places held out, many did not. On 2 October Jerusalem surrendered after a short siege. The Kingdom of Jerusalem, established by the First Crusade in 1099, had effectively ceased to exist.

Napoleon in Russia, 1812

In 1812 Napoleon was at the height of his powers. The central European states and empires had been conquered and many of their troops incorporated into the French army. In the west, the British, Portuguese and Spanish remained defiant, but Napoleon's biggest concern was to humble Russia. He amassed the Grande Armée, crossed the Vistula and sought out the Russian Imperial forces. Napoleon intended to bring them to battle and inflict a decisive defeat, just as he had done at Austerlitz in 1805 (see p. 81).

General Kutuzov and the Russian staff were in no hurry to confront the sprawling French army, and were happy to wait for a favourable opportunity for battle. They withdrew deeper into the interior of Russia until they reached Borodino, where they made a stand in order to protect Moscow. Drawing their forces up on low hills, with their batteries protected behind earthworks, they endured a series of frontal assaults from the French and the pounding of Napoleon's massed guns. The battle was indecisive: Napoleon failed to smash the Russians, and Kutuzov merely continued his withdrawal.

Napoleon decided to risk the further extension of his already overstretched supply lines and dash to Moscow, hoping that the fall of the city would compel the Tsar to seek terms. But when he arrived, the city was empty. Kutuzov had lured Napoleon into the heart of Russia but denied him a victory. When the French turned to withdraw westwards, the Russians began a series of harassing attacks. As temperatures fell and the system of supply broke down, French troops found themselves constantly raided by Cossacks and attacked by regular Russian formations. The arrival of the snows slowed their retreat and enabled the Russians to pick off isolated French detachments and stragglers. At the Berezina crossing, many French soldiers were drowned, killed in the rearguard action or captured.

The Retreat from Moscow: Napoleon was drawn into the interior of Russia and when forced to withdraw was defeated by Kutuzov and the harsh winter conditions.

The frozen remnants of the Grande Armée, a fraction of the original force (of 422,000, just 10,000 were combat effective at the end of the retreat), eventually re-crossed the Russian border. The Russians had carefully marshalled their resources and made full use of 'General Winter', the vast distances of Russian territory and the people's hatred of the foreign invader to defeat the most powerful army in Europe. The Retreat from Moscow was the turning point in Napoleon's fortunes, and his demise was due to the patient and calculated drawing of the enemy by General Kutuzov.

Analysis

Drawing the enemy into a predetermined killing area or into an area deprived of cover or supplies can sometimes be achieved by deception and ruses, but, in the case of Hattin and the invasion of Russia, the victorious commanders permitted their enemies to make mistakes and capitalized on them. Saladin and Kutuzov knew their enemy: they realized that their opponents could be induced to strike into 'thin air', but then deprived of the most vital supplies – water at Hattin and adequate rations in Russia. The same techniques might also be used to draw an enemy into an ambush, or on to the guns and missiles of a main body. As the examples here show, the boldness and strength of an enemy can be neutralized very effectively by this technique.

20

Kurikara, 1183
Q-Ships, 1915–17

Deception and Feints

Deception needs coherent planning, tight security and, more often than not, plenty of time. To 'work', the deception has only to establish significant doubt in the minds of the enemy so that they alter their original plans. Deception can be applied to both defence and attack, and can be useful in any mode of warfare, from conventional to irregular, in confrontations like the Cold War and in actual combat. Sun Tzu believed that its importance was so great that it had to be a feature of every conflict. However, as noted, modern era strategists such as Clausewitz felt that planning deceptions could be a distraction from the essential tasks of the commander and that surprise was difficult to achieve when intelligence and reconnaissance were comprehensive. He did, however, acknowledge that, in a state of weakness and powerlessness, 'cunning may well be the only hope'.[1] Jomini also felt that deception was a secondary activity compared with success at the decisive points.[2]

Ruses of war have been regarded, at different times and places, as illegitimate means of warfare, but that has never stopped their use. However, abuse of a truce or white flag, exploiting medical transport to manoeuvre troops, and donning the uniforms of the enemy to perpetrate some act of treachery, are universally condemned. On the other hand, disguising warships as merchant ships in order to catch a raider was an accepted stratagem used in the Napoleonic wars and in the two world wars.

20

Previous pages: Deception enables a commander to conceal his plans, his forces and his movements. These dummy invasion craft are moored in an English river before D-Day.

Surprise was crucial to the outcome of Kurikara in 1183.

Although elaborate deception operations are rare, the feint attack has been a common and effective stratagem in battle. It has the appearance of an attack, and even if the enemy commander suspects it may be a diversion, he still has to take it seriously and react to it. In a prepared defence, deception may be easier to achieve because there is more time. Colonel Baden-Powell, commander in the besieged town of Mafeking during the South African War (1899–1902), created dummy minefields and forts:[3] there were 'scarecrow' sentries, logs used to imitate cannon barrels and a 'searchlight' created out of a biscuit tin and a lamp. He also ordered his men to pretend to be crossing barbed-wire barriers at certain points. As a result, Baden-Powell's Afrikaner enemies overestimated the strength of the defences.

In Chinese military history, perhaps the most famous collection of methods of this kind was the *Thirty-Six Stratagems*, attributed to General Wang of the Southern Qi (late 5th century AD).[4] Phrased in proverbial terms, the techniques are still easily recognizable. They include: 'Entice the tiger to leave its lair [lure the enemy out of his position]; slough off the cicada's golden shell [mask one's forces and create an illusion]; and deck the tree with false blossoms [camouflage and deceive].'

Kurikara, 1183

Yoshinaka, a skilled commander of the Minamoto clan in the Japanese Gempei Wars (1180–85), used deception to win a decisive battle. In 1180 he conducted a scorched earth campaign against his long-term enemy, the Taira clan.

However, the raiding ceased during a two-year famine in his territories, giving the Taira the chance to recover and prepare a counter-attack. The Taira army, led by Koremori and Shigemori, was estimated at 40,000 men, with some accounts suggesting it was as large as 100,000. The sheer scale of the force indicates that the Taira sought a quick, decisive victory, but there were other reasons for haste. First, finding enough food in the Etchu region would be difficult. Secondly, many of the soldiers were recruited from neighbouring clans and a substantial proportion of these were militia, eager to return to their farms.

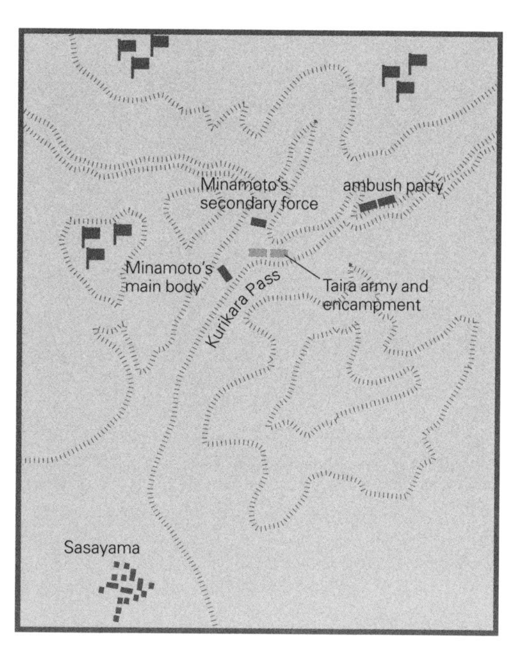

Yoshinaka's concealed dispositions and deception ruses, including flags on adjacent high ground to make it look like a large force was waiting in ambush, gave him a decisive victory over the Taira at Kurikara.

The Taira entered the Minamoto lands in two columns, one entered from the north, through Noto Province, while the other, led by Koremori, toiled up to the Kurikara Pass. As they approached, Yoshinaka ordered great numbers of white flags to be placed on an adjacent hill. These flags, the colour of the Minamoto clan, suggested that a large force was stationed on the flank of the pass to attack the Taira as they proceeded. Given the late hour that the Taira reached the crest, Yoshinaka calculated correctly that they would camp and resume their advance in the morning.

Using cover of darkness, Yoshinaka sent portions of his force to conceal themselves in the rear of the Taira army. One arranged itself in ambush along the likely line of retreat, down the pass. Another readied itself to strike from behind. The main army remained near the pass to confront the Taira under Yoshinaka's direct command. In the morning, Yoshinaka appeared to fight the battle the Taira had expected. After an exchange of arrows, including the unsettling 'whistling bulb' variety, Japanese warfare doctrines extolled individual combat. The literary exploits of heroic warriors provided inspiration and the Taira were eager to match these. So, instead of using their advantage in numbers, the Taira sent forward a handful of men who engaged in the rituals of hand-to-hand combat.

Yoshinaka used this time to bring up a herd of cattle. With flaming torches tied to their long horns, they were suddenly stampeded into the densely packed Taira troops. Hemmed in by the valley sides, escape was difficult. Several tried to counter-charge the animals in the hope of turning them, but were trampled. With the Taira thrown into confusion, Yoshinaka's men swept forwards. The whole Taira force started to give ground, and was then attacked from behind by Yoshinaka's second, concealed group. The shock was too much and the Taira started to flee back down the pass, to be ambushed by the third concealed force. The Taira were routed.

Q-Ships, 1915–17

During the first two years of the First World War, Britain's Royal Navy struggled to find a solution to U-boat attacks in the Atlantic. The German submarines were attacking merchant vessels, especially individual ships, thereby threatening to starve the British of food and war materials. It was noticed early on, however, that rather than stay submerged and deploy their torpedoes, the Germans preferred to surface and use the less sophisticated main armament on the foredeck to sink their victims. The Royal Navy could pursue a U-boat if it knew its approximate position, but depth charges were unreliable, so a submerged submarine was difficult to defeat; if it could catch the U-boat on the surface, however, it could bring its guns to bear or ram it.

The British plan was to deceive the U-boat commanders and encourage them to surface.[5] A number of civilian vessels were fitted with concealed armaments and given the code-name Q-ships (after their home port, Queenstown in Ireland). With holds packed with wood to enable them to float even when torpedoed, they were deliberately sent into areas where U-boats were known to be operating. If a U-boat was sighted, a mock 'abandon ship' drill could be staged by a pre-determined 'panic party'. As the U-boat surfaced, side panels on the Q-ship were dropped to clear the line of fire for concealed guns. Disguises for the ships grew more elaborate as the German commanders became alert to the threat.

Just off Great Yarmouth in Norfolk in August 1915, the converted trawler HMAS (His Majesty's Armed Smack) *Inverlyon* engaged *U-4*. The apparently innocuous sailing vessel was armed with a 3-pounder gun and small arms. As the U-boat surfaced, *Inverlyon* bared its teeth and poured fire into the hull and tower of the submarine at close quarters. Nine rounds sank the U-boat, with no survivors.

A different tactic had been used the previous June when *U-40* was sunk off Aberdeen. With the Q-ship *Taranaki* acting as bait, a Royal Navy submarine, HMS *C-24* fired on and destroyed the U-boat as it attempted to surface and finish off the *Taranaki*. Another Q-ship, HMS *Baralong*, sank *U-27* as it surfaced to attack a merchantman off southwest England. Some of the German submariners escaped and swam off, intending to seize their victim, the merchant ship. Desperate to prevent this, the *Baralong* opened fire, but many of the U-boat crew still reached the merchant vessel. They were counter-attacked by a boarding party from *Baralong* led by its commander, Lieutenant Godfrey Herbert. In the action, the Germans were wiped out. Commander

Crew manning a gun on the Q-ship *Brig*. Apparently vulnerable merchant vessels could conceal formidable armament to defeat U-boats.

Gordon Campbell commanded three Q-ships, sinking three U-boats off the coast of Ireland, and, with the *Dunraven*, damaging a fourth in an extended surface action in the Bay of Biscay.

Perhaps the most spectacular success was that of HMS *Prize* in April 1917. The *U-93* surfaced to attack this apparently defenceless ship and, approaching it, subjected it to heavy shellfire. For several minutes, the British endured a severe bombardment, but the captain of the *Prize*, Lieutenant Commander William Sanders, waited until the U-boat was within 80 yd (73 m) in order to have the best chance of success. Suddenly, the White Ensign of the Royal Navy was hoisted and the British ran out their guns. In the short exchange *U-93* was crippled and appeared to sink, although in fact it managed to limp back to port much later. Sanders was awarded the Victoria Cross, Britain's highest decoration for gallantry, for his leadership.

There were at least 150 engagements between U-boats and Q-ships, resulting in the destruction of 19 German vessels and damage to another 60. The Germans found no solution to the '*U-Boot-Falle*' (U-boat trap), although they managed to sink 24 of the 193 Q-ships that were commissioned once their security had been blown.

Analysis

Sun Tzu urged his readers to do the opposite of what is expected: 'Attack where he is unprepared; sally out when he does not expect you'.[6] And the *Thirty-Six Stratagems* recommends: 'Make a sound to the east, then strike in the west'. While all armies have sought to reduce their losses, deception is often employed by forces that are weaker, since powerful armies may neglect its use.[7] Nevertheless, even in strength Sun Tzu referred often to feigning weakness to draw the enemy on to favourable ground by means of false withdrawals, disordered ranks, or noises and sounds that could be observed. Moreover, the element of surprise was the result of the ability to *deceive* the enemy and concentrate forces where the enemy did not expect them – a situation achieved by Charles Martel at Amblève.

At Kurikara, Yoshinaka's clever use of flags, stampeding animals and concealed dispositions contributed directly to the Minamoto victory at the decisive point. The Taira thought they were conducting war on their terms, but Yoshinaka turned the terrain to his advantage. Hannibal was another commander to exploit the ruse of tying flaming torches to cattle, in order to extricate his army from entrapment in the mountains of central Italy in 217 BC. The British Q-ships lured the German U-boats into attacking as an effective way of protecting vulnerable allied shipping. In these examples, deception was a means to offset a significant weakness.

Accurate knowledge, as ever, is one key to military success, and deception aims to deprive the enemy of this. Reconnaissance, intelligence – perhaps through intercepted communications – and also espionage play their part, and deception may operate in this sphere through the use of double-agents to sow false information. In the Second World War the most celebrated example was Operation Fortitude, the deception campaign prior to D-Day in 1944.[8] The aim was to conceal the main point of attack by creating the false impression that the Allies would cross the Channel to the Pas de Calais. They devised the fictitious FUSAG (First United States Army Group) under General Patton, complete with dummy tanks, depots and radio traffic. Information deliberately 'leaked' to the German authorities seemed to confirm that the Pas de Calais was the objective.[9] To add realism, snippets of information about an attack in Normandy were released, but with the suggestion that these were diversionary attacks.

Operation Fortitude was a success, with German reserve formations held back from Normandy in the Pas de Calais area. The deception had a clear plan, the 'story' was believable, security was excellent, but the impression was given that the Allies were careless with their secrets. The 'Double-X' system of feeding double-agents with misleading intelligence worked smoothly. The deception itself was in direct support of an operation at the decisive point, and surprise enabled the Allies to concentrate a strong force where the enemy, who had been lured out of position, was not expecting it. As Sun Tzu concluded: 'All warfare is based on deception'.[10]

21

Thebes, 335 BC
Palestinian Terrorism, 1950–99

Terror and Psychological Warfare

Terror has frequently been used as an instrument of warfare, although its effectiveness is much disputed. Breaking an enemy's will to resist and frightening them from future activity because of the fear of grave consequences can be a key factor in securing victory. At a low level terror may be little more than harassment, reflecting the perpetrators' inability to harm their enemy more seriously. In the late 11th century, religious justification for the murder of enemy statesmen was used by the Muslim Shia Ismaili sect known as the Assassins. Led by Hasan-i Sabbah, their spree of killings caused considerable problems for both crusader and Muslim rulers in the Near East; to everyone's satisfaction they were wiped out by the Mongols in 1256.[1] For revolutionaries of the modern era, terror was regarded simply as a necessary tool of coercion – to conquer an oppressive state's will to govern, or to force the people to conform to the new ideology. Robespierre, the French revolutionary leader, wrote: 'They say terror is the resort of a despotic government. Is our government then despotism? Yes, as the sword that flashes in the hand of the hero of liberty is like that with which the satellites of tyranny are armed ... The government of the revolution is the despotism of liberty against tyranny.'[2]

Developments in weapons technology, for example reliable lightweight explosives, mercury detonators and long-range firearms in the mid-19th century, coupled with an increasingly urban population in the West, permitted modern forms of effective terrorism by small cells of dedicated activists. In Russia in 1877, the *Narodnaya Volyna* (People's Will) began a campaign of political violence against the apparatus of the state using explosives and firearms.

21

Previous pages: Palestinian fighters of Fatah in 1991. The aim of terror tactics is to keep resistance alive and break the enemy's political will.

Terror and psychological warfare have long been used on the battlefield. The Mongols laid waste to settlements that offered resistance in an attempt to persuade larger neighbouring populations to capitulate. In the Second World War, Japanese military training included not only the indoctrination of its own troops, but emphasized the shock effect of bayonet charges. In the Korean War, the Chinese and United Nations both exploited psychological warfare, with loudspeaker broadcasts across the front lines, leaflets dropped on to enemy positions and political indoctrination of prisoners.[3] Throughout history, states and armies have resorted to massacres and the sacking of settlements to intimidate civilian populations. Civilians and defeated armies have also turned to unconventional operations to sustain their resistance.

Thebes, 335 BC

Alexander charging into battle against Darius at Issus, where he achieved a great victory over the Persians. He also knew how to use intimidation to break the will of his enemies, as at Thebes.

Alexander the Great, in addition to being a brilliant tactician, proved adept at manipulating the emotions of friend and foe alike. On ascending the Macedonian throne in 336 BC as a youth of 20, he was faced by the need to stamp his authority on both his court and the Greek world: while his father, Philip, had established a powerful reputation through two decades of successful campaigning and effective diplomacy, his own achievements could be disparaged by enemies. A particular problem was posed by the Greek cities, especially the most powerful, Athens and Thebes, whom Philip had cowed into acquiescing in the Macedonian-led League of Corinth after defeating them at Chaeronea (338 BC).

Alexander inherited leadership of this anti-Persian League, and a rapid advance into Greece in 336 BC had persuaded the League members to endorse his authority, pre-empting possible action by anti-Macedonian groups in Athens and Thebes. As part of his preparations for invading Persia, Alexander next turned to the northern and western borders of his

kingdom to assert his leadership there. In 335 BC, his lengthening absence led to rumours of his death, prompting the Thebans to rise against the Macedonian garrison occupying the Cadmea, their citadel; they also appealed to the Athenians to join in liberating Greece from Macedonian tyranny.

Alexander's response was immediate. Two weeks after learning of the Theban revolt, he had marched his army across 240 miles (386 km) of rough mountains to arrive in Boeotia at the same time as the news of his approach; his sudden appearance prompted troops marching to support Thebes from the Peloponnese to turn back, while the Athenians had sent military supplies but not yet troops. Although the details of Alexander's attack on Thebes are disputed, it seems that after hard fighting the Macedonians were able to force their way into the city and join up with their garrison in the citadel; as Theban resistance cracked, there was widespread slaughter of the fleeing soldiers. Alexander's exploitation of the capture was equally decisive. The decision was taken to raze the city to the ground, with the exception of temples and the house of the 5th-century poet Pindar, who was famous for celebrating the victories of Greek aristocrats, and to enslave the entire population, except for priests and priestesses, friends of Macedonia and Pindar's descendants. Apologists for Alexander, in the ancient world as well as more recently, blame this decision on Alexander's Greek allies, especially Thebes' neighbours in Boeotia who were keen to eliminate their local rival, but the action could not have been taken unless it coincided with Alexander's own wishes.

There were precedents for the ruthless treatment of captured cities in the Greek world: Alexander's father, Philip, had destroyed Olynthus and enslaved its inhabitants, while Thebes itself, after capturing Orchomenos, had slaughtered the males and enslaved the remaining population. But Alexander's treatment of Thebes stood out, partly because it was one of the major cities of the Greek world and so would have seemed immune to such a fate, and partly because of the rapid reversal from confident and insulting opposition to Macedonian power to complete annihilation. This was the intention. Alexander, preparing to leave Europe to launch his Persian campaign, will have been well aware that the traditional Persian response to such challenges was to deploy their 'golden archers' – the gold darics which were stamped with an image of an archer – to purchase support among the Greek cities and so foment trouble back home for invaders.

Alexander's strategy worked. In 334/3 BC as his invasion force advanced across Asia Minor, the Persian navy under Pharnabazus campaigned effectively in the Aegean, gradually moving towards the Hellespont, the direct communication route between Alexander's army and its Macedonian home, as well as the channel through which sailed part of the grain supply of Athens and other cities. However, in spite of presenting such an opportunity, Pharnabazus received no support from Athens or other leading states fretting under Macedonian domination. In the case of Athens, the fear instilled by the destruction of Thebes was reinforced by concerns about the fate of Athenians whom Alexander had captured in the Persian army at the Granicus (334 BC), as well as of 20 ships with their crews of about 4,000 rowers which Alexander had deliberately retained when disbanding his navy in 334.

Palestinian Terrorism, 1950–99

The use of terror by Palestinians in the modern era can be traced to the 1920s and 1930s, when the immigration of Jews from Europe conflicted with local Arab interests. After the Second World War, the Irgun terror organization, established by Jewish settlers, set out to disrupt the British mandate forces in Palestine and hasten their withdrawal – thus making way for the establishment of a state of Israel. In spite of some spectacular attacks, such as the bombing of the King David hotel in Jerusalem, British curfews and other counter-measures curtailed their effectiveness and Irgun terror contributed little to the British decision to withdraw in 1948.

When Palestinians were forced from their homeland by the subsequent Arab-Jewish fighting in 1948, a whole generation of young men sought revenge for *Al Nakba* ('the disaster'). They could not defeat the Israelis, but relied on infiltrations by small groups of *fedayeen* ('freedom fighters' or 'martyrs'); in 1950 these caused the deaths of over 100 Israeli civilians and this steady toll continued through the decade. Attacks were also made on farms and transport systems. The Israelis responded in a manner that subsequently became standard: counter-attacks, raids and targeted assassinations.

In the 1950s Fatah ('Victory'), a Palestinian resistance organization, was established. Using funds from Saudi Arabia, Fatah trained and indoctrinated its fighters, and launched scores of small raids from Jordan, Syria, Gaza and Lebanon. Although the overall aim was the 'liberation' of Palestine, the raids were chiefly to keep resistance alive. Yasser Arafat, who emerged as a leader who could mobilize the Palestinians, worked hard to persuade Arab states to fight for the Palestinian cause, but the 1967 war weakened the Palestinian position and they were divided into a multitude of competing groups. Some co-ordination was achieved in 1969 when Arafat became chairman of the Palestine Liberation Organization (PLO), an overarching resistance body.

From the late 1960s, the PLO tried to build up public support by providing schools, hospitals and training centres. Arafat had no illusions that to form a state would simply invite an Israeli attack, but without proper organization, his PLO followers were often ill-disciplined. While based in Jordan, PLO fighters frequently terrorized and attacked the civilian population. In Black September (1970–71), Arafat pushed his hosts too far and fighting broke out with the Jordanian army. He and his supporters were expelled to Lebanon. To raise the stakes against Israel, Arafat sanctioned the attempt to kidnap Israeli athletes at the Munich Olympics in 1972. The mission went awry: five Fatah kidnappers, a police

The Munich Olympics, 1972: a small terrorist cell demonstrates its disproportionate impact.

officer and 11 Israeli athletes were killed in a counter-terror operation. Against a background of continued PLO outrages on Israeli civilians, the Israeli Defence Force (IDF) stepped up its counter-attacks and cleared more Palestinians from the West Bank. The Munich Olympics operation failed to provide Arafat with any international sympathy, but the results further embittered his own followers against Israel and therefore deepened their commitment to the cause.

By the end of the first Lebanon Civil War in 1977 the PLO were in control of the southern region, bordering Israel, and they began to reconstruct facilities such as schools and clinics to present themselves as a state-worthy organization. Israel wanted to see the PLO defeated and raided across the border. In March 1978, a PLO gang retaliated by infiltrating Israel, where it murdered 37 civilians and wounded 78 more during an ambush of two buses. The first Israeli response was to clear the PLO from a 'security zone' south of the River Litani, but in 1982 they used the attempted murder of their ambassador to Great Britain by the Abu Nidal group as a pretext for a full-scale invasion of Lebanon. Again the PLO was defeated in conventional operations, and found itself bottled up in central Beirut, but still it survived, with Arafat's perennial optimism infuriating his enemy.

The Intifada uprising of 1987 initially boosted Arafat's reputation, but harsh Israeli counter-measures spawned more radical groups that developed as rivals to Fatah, most prominently the Islamic Resistance Movement, later titled Hamas. These Jihadist groups rejected any idea of negotiation with Israel and promoted unlimited war. International opinion began to sympathize with the stone-throwing Palestinians and turn against the 'heavy-handed' response of Israel. In radical circles, every death was consecrated as a 'martyrdom'. Hamas and Islamic Jihad resumed their offensive of missile attacks, bombings and shootings as a second phase of the Intifada broke out in 1992. Their delivery of effective challenges to Israel undermined Arafat's dominance of Palestinian politics, gradually turning him and Fatah into forces for moderation in spite of their continued support for terrorist activities when the opportunity presented.

Analysis

Alexander's terror at Thebes was the means to intimidate decision-makers in Greek states, primarily Athens, thus permitting him to concentrate on operations against Persia. Intimidation has often been used to break the will to resist, and may manifest itself in the weapons, appearance and tactics of armies – dive-bombers fitted with sirens is one example. For modern terrorists, the strategy may be as much to cause systemic disruption as to destroy targets. Terror can create a disproportionate effect on the delicate and vulnerable threads of society, but its use by underground groups cannot, of itself, secure power. Such groups seek to discredit governments by demonstrating that the authorities are powerless to protect the public, and therefore that only a change of policy will end the conflict. Arafat and the PLO, by sustaining an armed struggle despite their obvious military weakness relative to Israel, kept Palestine on the international agenda, and, by provoking strong reprisals, earnt some sympathy around the world. Nevertheless, more successful terrorist actions might undermine this achievement, since the indiscriminate killing of civilians hands the propaganda initiative back to the state and invites heavier counter-attacks.

Terror is a complex weapon, the results of which are diverse. It can prompt greater determination, as during the London Blitz in the Second World War, generate sympathy for the underdog, as in much of the Intifada, or alienate potential supporters, as in the Omagh bombing by the Real IRA in Northern Ireland in 1998. Yet the PLO strategy was consistent in its aim, even if it faced many setbacks: to keep alive Palestinian resistance. Arafat lost the initiative and clearly made tactical errors. His refusal to accept diplomatic settlements were lost opportunities, but he believed that the unity of his fragile movement was a greater priority. The PLO and its rival organizations used every opportunity to attack Israelis in an attempt to change the government's policy. While terrorism provoked Israeli counter-attacks, the PLO tried to turn this into a means of discrediting Israel or at least highlighting the iniquities of the 'occupation'. Terrorism thus combines physical assaults with an attempt to break the will or mind of the enemy, or, at the very least, to play out the drama of one's cause on the grand stage of world opinion.

22

Verdun, 1916

Attrition and Annihilation

Not all battles are won by manoeuvre. If the centre of gravity lies in the strength of the enemy army, then an effective tactic is the destruction of that force, as with the German Panzerarmee at El Alamein (see **1**). When a decisive victory is impossible, because of the strength of the opposing positions for example, then the enemy forces must be worn down physically and psychologically to a point where they no longer possess the will or the capacity to fight. One aim of attrition is brutally simple: to kill as many of the enemy as possible, a strategy of annihilation. Attrition has often been associated with modern industrialized wars because rapid-firing precision weapons can effect mass slaughter, but parity of forces, as in the American Civil War or the First World War, has also been a cause: unable to achieve a decisive result, the armies bludgeoned each other over extended periods, until one was too weak to continue.

In the trench warfare of the Western Front in the First World War, the British developed the idea of 'bite and hold', that is, taking a portion of the enemy line and forcing the Germans to make costly counter-attacks. The emphasis here was on inflicting casualties.

Previous pages: The devastated battlefield of Verdun, where chances of survival for the infantry seemed bleak. Here a French light machine gun crew shelter in a shell crater near Fort Vaux.

The notion of 'wearing out' the enemy, however, is not limited to mass bloodletting. Against Hannibal, the strategy of Fabius Maximus Cunctator, 'the Delayer', was to wear down his formidable opponent by avoiding the pitched battles in which the Carthaginian had been victorious. Similarly, Hans Delbrück, the German military historian, suggested that *Ermattungsstrategie*, 'wearing out strategy', could be achieved by manoeuvre or sieges, without costly battles;[1] he argued that Frederick the Great avoided action on unfavourable terms and manoeuvred to minimize losses. Thus, for Delbrück, attrition was a form of manoeuvre and to be distinguished from *Niederwerfungsstrategie*, the strategy of annihilation.

Clausewitz, in contrast, did not advocate attrition as an operational concept, though the decisive destruction of the enemy army was a central idea for him. Stalemate and mutual attrition could threaten the ability to achieve political objectives, and instead gave supremacy to continuing war, with tactical considerations displacing strategy. This undermined the rational calculus of war. This risk is well illustrated by the Battle of Verdun.

Verdun, 1916

During 1915 several attempts were made to break the stalemate of trench warfare that had developed on the Western Front by the end of 1914, but British and French attacks and German experiments with poison gas at Ypres produced no decisive results. The Germans were outnumbered, but were well organized and had developed strong defences. On the other hand, the Allies were hampered by insufficient equipment, a shortage of heavy weapons and ammunition, and poor communication facilities, and had no tactical solution to avoid the heavy casualties involved in attacking across no-man's land.

Nevertheless, both sides believed that 1916 would be the turning point. Germany could now spare troops from the Eastern Front where it had temporarily checked Russia, while Britain and France expected to be able to bring to bear larger and better-equipped forces as their war industries geared up. Thus the year 1916 had to be the year of the 'big push'. Germany planned to break the trench stalemate at Verdun by wearing down the French army, but the Allied commanders also believed that they would see an inevitable consequence of static and industrialized warfare that 'wore down the strength of the German Armies'.[2]

The designer of the German plan of attack was General Erich von Falkenhayn. He calculated that if France could be knocked out of the war, Britain would lack the will to continue alone. Falkenhayn therefore selected a target of such importance to the French that they would deploy every man in its defence, where they could be destroyed by artillery fire. The French would thus be forced in effect to sacrifice their troops; Falkenhayn apparently referred to this as *weissbluten* ('to bleed [the French] white').[3] Verdun was selected because its ring of fortresses had a national significance. It was a garrison-town situated on the River Meuse in the Region Fortifiée de Verdun (RFV) and surrounded by a double circle of 20 large forts and 40 medium-sized fortifications in hilly country covered with woods and criss-crossed with deep clefts and gorges. The French line was about 8 km (5 miles) from Verdun itself.

Operation Gericht, the German plan to reduce this complex, was set for February 1916. The German Verdun sector was under the nominal command of Crown Prince Wilhelm, but his Chief of Staff, General von Knobelsdorf, took the key decisions. Knobelsdorf thought the aim was to capture Verdun itself, whereas Falkenhayn's orders spoke of 'an offensive in the surroundings of the River Meuse, in the direction of Verdun'. Although Falkenhayn's plan was to destroy successive French units as each was replaced using massive artillery bombardment, he also believed that his own troops would be motivated better by an attack on a specific objective than the concept of attrition: he allowed his commanders to lose sight of the original plan.

The shelling began over the entire Verdun sector along a front of 25 miles (40 km) on 21 February. The nine-hour bombardment was the heaviest of the war to that date: trenches caved in, men were buried alive, and key choke points in the French lines, identified before the battle, were subjected to an intense barrage. The devastation was appalling, with entire trench systems, phone lines and artillery destroyed. Men were torn to pieces by shells and the

22

The French 'Voie Sacrée': a crucial supply line that enabled them to match and then annihilate German forces at Verdun.

psychological effect on survivors was extreme.[4] By 25 February, the French were being forced to withdraw back towards the Meuse. General de Castelnau, arriving at Verdun, believed that an evacuation would constitute a serious blow to French morale. Just as Falkenhayn had calculated, Verdun was now elevated to a symbolic status out of all proportion to its strategic value. De Castelnau ordered the right bank of the river to be held at all costs. Fresh troops were brought in and he appointed General Philippe Pétain to defend the approaches to the city.

Pétain formed a line between the remaining fortresses (which were immediately reinforced), to be maintained at all costs, to permit the arrival of reinforcements from his own Second Army. But Pétain understood that this was an artillery battle: an increase in French guns would halt the German attacks by cutting down their men. He therefore organized a supply route to bring up the great quantities of ammunition his batteries would require. There was just one narrow connecting road between Verdun and Bar-le-Duc, but he deployed large numbers of men with picks and shovels day and night to keep open this life line, the 'Voie Sacrée'.

Turning the Tide

Pétain now called upon his men to make an epic defence. The battle for Douaumont was typical. The village was situated right next to a fort which had already been captured by the Germans, and was now itself turned into a fortress, with a fresh French garrison, more than 30 machine gun nests and support from a battery of howitzers. French fire was highly effective in the open ground in front of the village because every scrap of cover had been swept away; German attacks on 26 and 27 February were severely mauled. As successive waves of infantry went in, the French maintained heavy fire, inflicting still greater losses.[5] Here was the chief problem for the Germans: after their initial successes, their own infantry was being drawn into a battle of attrition as the French adapted the German plan and turned the tables. Although German guns were still killing large numbers of French soldiers as they struggled to man defences or to reinforce the battered lines, the Germans too were suffering severely.

22

Having checked the German offensive, the French inflicted heavy losses and retook lost ground through 1916.

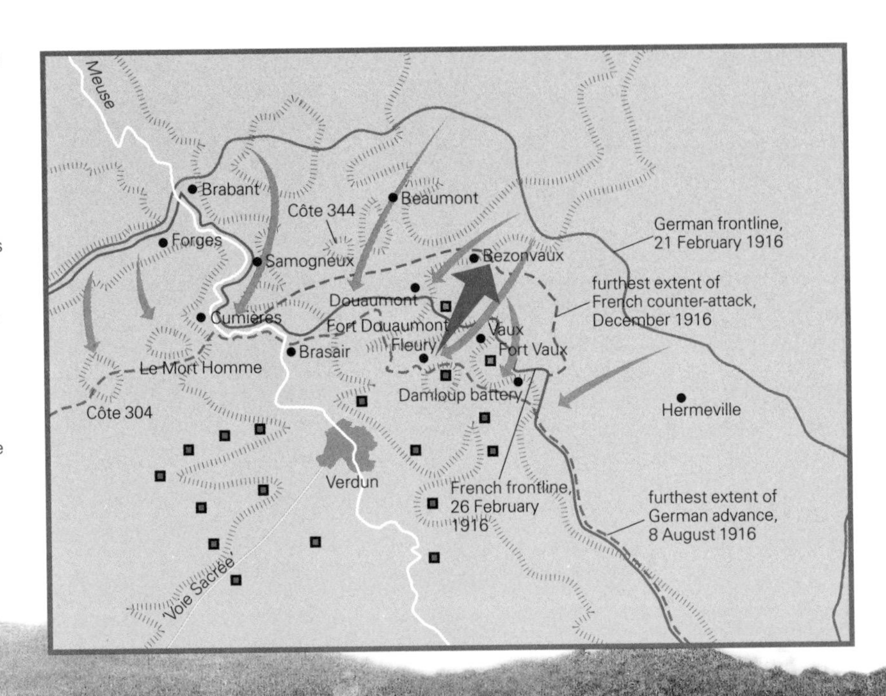

A German dugout is converted into French trenches after the capture of Fort Douaumont. Even trenches provided little protection: Verdun was the 'Mill on the Meuse' that consumed men.

Under Pétain's command, the French artillery had increased to 500 guns, concentrated on the Meuse's left bank. From here, protected by the hills and ravines, and beyond reach of German shells, the guns had a devastating impact as the French began to rake each valley systematically. An attack by the German VII Corps on 27 February 1916 was checked as it tried to cross the Meuse to knock out these French batteries. It was proving impossible to advance in the open without suffering heavy casualties. German soldiers were now experiencing the same psychological pressure as the French, and noted the omnipresence of death: 'We all carried the smell of dead bodies with us. The bread we ate, the stagnant water we drank. Everything we touched smelled of decomposition due to the fact that the earth surrounding us was packed with dead bodies.'[6] By the beginning of March, German commanders concluded that French artillery had to be neutralized before Verdun could be successfully attacked.

However, the Germans were still drawn to individual objectives. Fort Vaux, a medium-sized fort on a ridge blocking the German advance, was defended by 600 French under Major Sylvain-Eugène Raynal. It was subjected to intense shelling and Raynal counted an average of 1,500 to 2,000 hits an hour at one point. Gradually the French were reduced in number until just 26 survivors, including Raynal, fought on behind a barricade of sandbags deep inside the fort. When their water and ammunition were exhausted, they were forced to capitulate, but they had proved that, although the modern battlefield was dominated by artillery, human endeavour still counted.

In late June, the French army was almost thrown back from the right bank of the Meuse for a second time, but concentrated artillery fire again thwarted a German breakthrough. General Nivelle stiffened the defence with a stirring order of the day: he praised the courage and endurance of the French troops, concluding with the demand: *Ils ne passeront pas!* ('They shall not pass!').[7] The morale of the German army was now badly affected by the immense losses. An infantry officer recorded: 'the soldiers were numbed by the sight of bodies without heads, without legs, shot through the belly, with blown away foreheads, with holes in their chests, hardly recognizable, pale and dirty in the thick yellow brown mud which covers the battlefield.'[8] Gradually, the German offensive lost momentum; by September it had ceased altogether. But the French were now in a position to attempt to recover their original positions and to inflict further losses on the Germans. Between 21 October and 19 December they began a systematic counter-attack.

The right bank of the Meuse was transformed into an enormous construction site: roads and artillery fortifications were built and ammunition was stockpiled. Fresh French divisions were trained on a mock battlefield complete with forts. The artillery rehearsed the technique of a creeping barrage to permit assaulting troops and shellfire to reach enemy lines almost simultaneously so that defenders had no time to react. Nivelle also deployed 400-mm Creusot-Schneider guns, which had an enormous penetrative power – they crashed through the concrete roof of Douaumont, forcing the Germans to abandon the position temporarily. During this French counter-attack, 240,000 shells were fired each day, twice the daily average of the battle hitherto. On 24 October, under this hail of fire, Douaumont was retaken, and on 3 November so were the ruins of Fort Vaux.

Gradually, each battered village on this moonscape of a battlefield was retaken, so that by the close of the counter-offensive, the French had recovered almost all the ground they had lost. They had fired an estimated 23,000,000 shells during the battle and inflicted almost as many casualties as they had lost: 377,231 French and colonial forces were killed, wounded or missing; the Germans lost 337,000. The French had endured the German offensive and turned the policy of attrition on their assailants.

The Verdun landscape was itself purged by the fighting. Nine villages were destroyed and have never been rebuilt; a large area was so toxic and dangerous because of unexploded shells and gas it was designated a *zone rouge* – it has never been cleared. Brooding over the battlefield today is the imposing ossuary containing the skeletons of 130,000 bodies that could not be identified after the battle. Small apertures in the building allow visitors a macabre glimpse of the consequences of the most dreadful strategy of war.[9]

Analysis

'The Mill on the Meuse' was a particularly severe battle of attrition because of the relatively narrow and confined front, the dense concentration of artillery and the difficulties in making any large-scale manoeuvre. In the absence of armoured formations, and with limited air power, the infantry was the only force that could take and hold ground. Even the strongest fortifications, such as the concrete casemates of Douaumont and Vaux, could be smashed by the heaviest guns: there were no safe refuges.

The Battle of Verdun was a 'wearing out' battle *and* a battle of annihilation. Apart from the casualties it sapped the resolve of the French *poilus* (infantrymen): in 1917 there was a serious mutiny on the Chemin des Dames which halted offensive operations. However, the Germans were equally shaken: the losses they suffered at Verdun and the Somme in 1916, then in the battles of 1917, and finally the bitter setbacks after spring 1918 (see p.114) led sections of the German army to refuse to fight on. While the Allies, now reinforced by the Americans, could continue to contemplate a strategic victory in 1918, German commanders could see no means to achieve their war aims.

23

The Battle of the Atlantic, 1941–45
Cape Matapan, 1941
North Cape, 1943

Intelligence and Reconnaissance

Battlefield victory can hinge on accurate and timely intelligence about the enemy, particularly his intentions and his capabilities. Polybius, the Roman historian, advised a general to 'apply himself to learn the inclinations and character of his adversary'.[1] According to Sun Tzu: 'He who knows the enemy and himself will never in a hundred battles be at risk.'[2] At the tactical level, most commanders have relied on immediate observations to discern the enemy's objectives. It is a well-known military maxim that 'time spent in reconnaissance is seldom wasted'.

Reconnaissance is the surveillance process by which crucial battlefield information is gathered on enemy strengths and dispositions, which, when analysed, may indicate the actions the enemy is likely to take. Intelligence also operates at a more strategic level, and extends from conditions relevant to a particular battlefield, for example environment, topography and weather patterns, to information about an enemy's political, military or industrial deliberations. Battlefield or 'tactical' intelligence may be gathered through small patrols on foot or mounted, or the interrogation of prisoners of war or local civilians; air reconnaissance was introduced when a balloon was sent aloft at the Battle of Fleurus in 1794, and developed in the 20th century with the evolution of specialist aircraft. Deeper behind enemy lines, special agents engage in espionage or sabotage. To these must be added the interception of enemy communications, or SIGINT (Signals Intelligence). All information, however gained, requires verification, a principle recognized by Machiavelli in the 16th century[3] and Baron de Jomini in the 19th: Jomini recommended that commanders form 'hypotheses of probabilities', or a logical and systematic analysis of the available information.[4]

23

Previous pages: HMS *Prince of Wales* meets an Atlantic convoy. The Royal Navy dominated the Battle of the Atlantic, using its intelligence assets to great effect.

Surveillance technology is becoming increasingly sophisticated, including unmanned aircraft, thermal imagery, radar, satellite photography and specialist listening devices. In the 2003 invasion of Iraq, the Americans digitally mapped all friendly and hostile units in a 'real time' electronic environment, 'seeing' deep into the Iraqi lines and tracking their movements.[5] Nevertheless, they found that in the counter-insurgency phase, local human intelligence and the 'mark 1 eyeball' of foot patrols were just as valuable as signals intercepts and satellite data.

The Battle of the Atlantic, 1941–45

In the Battle of the Atlantic, codebreaking enabled the Allies to locate and defeat German U-boats, thereby turning the tide. Building on their experiences in the First World War, the British had created the GC&CS (Government Code and Cipher School), which was expanded just before the outbreak of the Second World War in 1939. A radio listening station was established at Bletchley Park with the code-name Station X (because it was the 10th station to be established), and the teams, numbering 9,000, were located there in wooden huts.[6] Many of the codebreakers were civilians recruited for their ability in crossword puzzles, mathematics and chess. A great deal of the work involved processing large amounts of encoded messages, many of them transmitted by German teleprinters known as 'Fish'.

The Battle of the Atlantic consisted of a struggle on both sides to produce new shipping or submarines to replace losses, a contest for air superiority and a technological race. In the early stages, the Germans had the upper hand, partly through their acquisition of the French Atlantic ports, partly because their naval intelligence, the B-Dienst signals service, had some success in breaking British codes so that Wolf Packs (U-boats operating in groups) could locate convoys of merchant shipping more accurately. In May 1941, after months of heavy losses, the British achieved an intelligence breakthrough: they managed to read intercepted German signals and consequently divert convoys away from areas where U-boats were operating. Moreover, U-boat supply vessels could be located and destroyed. Until the Germans altered their codes in February 1942 the Allies had the upper hand.

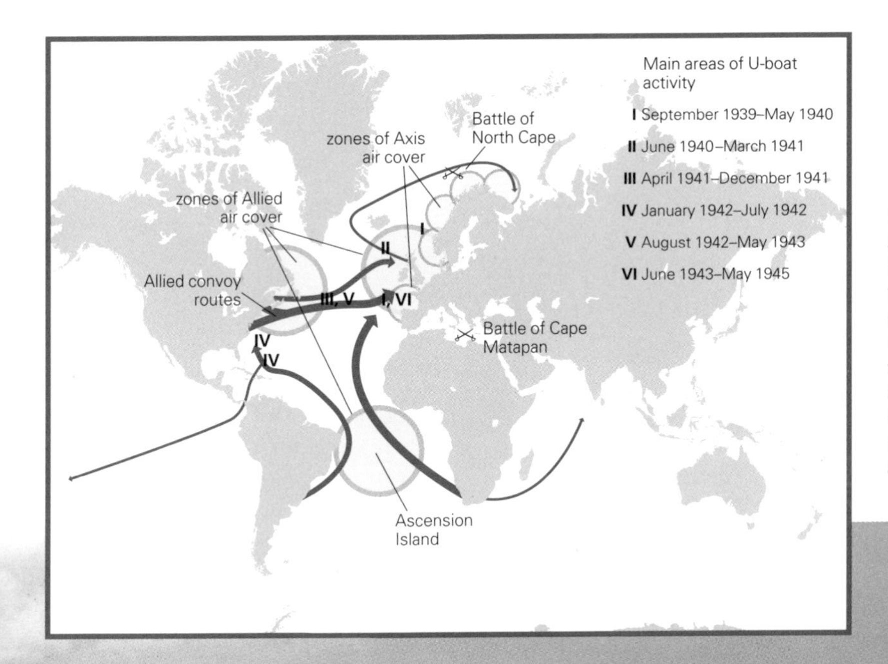

In the Battle of the Atlantic, breaking German ciphers was crucial to protect the merchant convoys, which kept Britain's supply lines moving, from German U-boat activity.

When the United States entered the war in December 1941, the Germans regained a brief advantage since the Americans did not adopt appropriate security measures until mid-1942, but thereafter U-boat successes diminished significantly and they were only able to operate without heavy losses off Brazil and southern Africa. The increase in the number of Allied support groups (escorts that could pursue U-boats), better detection and depth-charge technologies, together with greater air power through deployments by new American carriers to provide air cover in the mid-Atlantic 'gap', again shifted the conflict in favour of the Allies.

There was also a significant intelligence landmark in December 1942: German codes were broken and remained insecure despite further revisions of their procedures and systems in March 1943. The Allies could now predict with great accuracy the location of individual U-boats, and so hunt down and destroy them faster than losses could be replaced. To maintain an effective counter-intelligence screen, Allied codes were upgraded in June 1943, frustrating the efforts of German Naval Intelligence. Despite a crisis for the Allies in March 1943, when a large tonnage of merchant shipping was lost, the U-boats suffered catastrophic losses in April; in May the U-boats 'withdrew' from the Atlantic.

From 1943, the British could also make use of a specially designed computer, known as Colossus, to crack the German teleprinter cipher called 'Tunny'. Moreover, specialists worked on and broke Japanese naval ciphers which enabled Allied forces to intercept and destroy the bulk of Japanese merchant shipping by 1945. However, the most famous achievement was the cracking of the Enigma cipher machine, a development assisted by the admirable theoretical work done by Polish codebreakers at the start of the war and by the capture of German machines from weather ships and a U-boat. By 1944, over 4,000 German messages were being decrypted every day by Allied codebreakers.[7] The Germans continued to send individual U-boats on token missions, but their own technological improvements in 1945 came too late to influence the outcome of the war.

Cape Matapan, 1941

The value of codebreaking was not limited to the Atlantic campaign. In March 1941, British Naval Intelligence learnt that a strong Italian fleet, consisting of a battleship, *Vittorio Veneto,* six heavy cruisers, two light cruisers and 17 destroyers, had set out to attack a British convoy in the eastern Mediterranean. This accurate report of the Italian deployment contrasted with that possessed by the Italians. They believed the British Mediterranean Fleet had only one battleship, when there were three; they were also unaware that the Royal Navy had an aircraft carrier in the theatre. However, in order to protect the source of this intercept intelligence, code-named ULTRA, the fleet commander, Admiral Andrew Cunningham, despatched a reconnaissance aircraft on an apparently routine patrol to the precise location of the Italian fleet. At the same time, Cunningham ordered the cruisers in Greek waters, *Ajax, Gloucester, Orion* and *Perth,* to rendezvous south of Crete with Cunningham's own flotilla from Alexandria, consisting of the carrier *Formidable* and the warships *Warspite, Barham* and *Valiant*.

On 28 March, the leading Italian cruisers sighted the British ships proceeding towards Crete and assumed they were trying to escape. There were short encounters but no serious engagement. The British rendezvous was achieved and Cunningham deployed two flights of torpedo bombers from *Formidable*. They caused profound shock and persuaded Admiral Iachino, the Italian commander, to abandon pursuit of the British cruisers and seek the protection of his own air force closer to the Italian coast. However, one torpedo struck the stern of the *Vittorio Veneto* and caused

The might of a naval broadside from HMS *Valiant*, directed on to its target by clear intelligence.

considerable delay. Cunningham seized the opportunity to press the Italian force further: a third wave of aerial attack crippled the cruiser *Pola*. In a fateful decision, the Italian commander sent a strong escort back to accompany the stricken vessel, while the rest of the fleet continued its withdrawal. As darkness fell, the Italians left behind were effectively blinded. They possessed no radar for tracking surface threats and their naval doctrine dictated that they should not engage in night actions. As a result, the British vessels, which had located the Italian ships by radar at 2200 hrs, closed with them undetected.

At just 3,800 yd (3,500 m), the three British battleships opened fire. Searchlights illuminated the Italian ships, and at such a short range the Royal Navy gunners could not miss. In just three minutes, two Italian heavy cruisers, *Fiume* and *Zara*, were aflame and out of action. Five minutes into the action and two Italian destroyers were sinking. Another destroyer, the *Gioberti*, was severely damaged and tried to escape. The weight of fire was overwhelming, and just one Italian destroyer escaped. The only resistance offered was some machine gun fire and the launching of a few torpedoes. The Italians had clearly been taken by surprise and the speed of their destruction simply gave them no time to react. Intelligence and reconnaissance had allowed the British to determine exactly when the Italians would be at a particular location and in what strength. A combination of radar, firepower and bold leadership had secured victory. The Italian navy was deeply shocked by the action, and did not venture out to sea in significant strength again. British mastery of the Mediterranean could now be challenged only through air and submarine battles

North Cape, 1943

In late 1943, the German navy, the Kriegsmarine, sent a powerful force to intercept the British Arctic convoy JW 55B, which consisted of 19 cargo ships accompanied by eight destroyers, as it conveyed vital supplies to Russia. The convoy's position and heading were reported by Luftwaffe aircraft to the German squadron of five destroyers led by the battlecruiser *Scharnhorst*, which then made best speed through wintry seas on Christmas Day to close with the British. However, the Germans were unaware that behind the convoy were two strong Royal Navy flotillas, alerted to the enemy approach by codebreakers at Bletchley Park. Force 1 was made up of three cruisers, while Force 2 contained a battleship, *The Duke of York,* a cruiser, and five destroyers. The Germans were sailing into a trap.

On 26 December, Rear Admiral Bey on *Scharnhorst* could not locate the British convoy. The weather conditions were poor, with heavy seas and bad visibility. To widen his search, he dispatched his five destroyers southwards, but the convoy had swung northwards, anticipating an attack, and was now clear.

Scharnhorst was alone when the British cruisers of Force 1 detected it at 0900 hrs. They opened fire at a range of 13,000 yd (12,000 m), and, despite the worsening weather conditions, scored two hits – one of which destroyed the *Scharnhorst*'s radar. The German gunners tried to reply with their own salvoes, but, without radar, they could only align their guns towards the muzzle flashes of the cruiser HMS *Norfolk* – the other two British ships were using a new propellant that reduced muzzle flash. In a thickening snowstorm, Admiral Bey believed he was outgunned and turned southwards to link up with his destroyer escort. As he sailed south, he ordered them on to co-ordinates for the convoy provided by a U-boat. Bey still hoped to sink the convoy and perhaps lure the protecting British cruisers away, but when the German destroyers arrived at the specified location, the convoy had long since escaped.

As *Scharnhorst* ran south, only HMS *Belfast* managed to keep it on the radar, but *The Duke of York* and the destroyers of Force 2 arrived within torpedo range. At 1648 hrs, *Belfast* illuminated *Scharnhorst with* starshells to give *The Duke of York* a clearer target. Its first rounds destroyed *Scharnhorst*'s foremost gun turret, and ignited its aircraft hangar. Bey steamed away north at high speed, to be engaged by *Belfast* and *Norfolk*, and although the *Scharnhorst*'s guns twice hit *The Duke of York*, the British battleship's gunnery remained excellent. One shell smashed through the *Scharnhorst*'s armour belt and exploded in a boiler room, dramatically reducing its speed and preventing escape. In spite of more

Scharnhorst was located and destroyed by *The Duke of York* in an intelligence-led operation on the North Cape.

hits from torpedoes, *Scharnhorst* grimly fought on until, at 1945 hrs, the ship capsized and sank, taking most of the 1,968 crewmen with it: only 36 survivors were rescued from the freezing waters.

A combination of sound intelligence and the chance destruction of the German radar, added to the appalling weather, fatally disadvantaged the Germans in this encounter. Advance intelligence of the German departure from Norway permitted the Royal Navy to concentrate greater force on the *Scharnhorst*, and radar then allowed its moves to be tracked closely.

Analysis

Acquiring tactical information on the battlefield is crucial in order to locate and destroy enemy units. Sun Tzu wrote that intelligence was a critical element in preparing for conflict, and it did not diminish in importance once fighting had broken out. He advocated what would today be understood as a 'net intelligence assessment' of topography, the enemy commanders and their forces' strengths. He advised on the signs to look out for in an enemy on campaign: 'when his troops lean on their weapons, they are famished; when drawers of water drink before carrying it to camp, his troops are suffering from thirst'.[8] He advocated a study of maps and distances, the maximum use of spies, and urged generals to make probing attacks: 'agitate him and ascertain the pattern of his movement ... learn where his strength is abundant and where deficient'.[9] By contrast Clausewitz emphasized the unreliability and inaccuracy of much military intelligence, hinting at several aspects of the problem as he saw it: the difficulty in interpretation, the sheer volume of information that can overwhelm commanders and the effect of the 'friction' of war (unexpected problems and uncertainty).

There has probably never been a 'full' intelligence picture for any army, and war is still the realm of chance, not least because decisions are made by human beings. Counter-intelligence and tight security limit the flow of information. However, in the Royal Navy's campaign in the Atlantic, against the Italians in the Mediterranean and in pursuit of the *Scharnhorst*, intelligence and reconnaissance contributed substantially to the victories.

24
China, 1934–49
Vietnam, 1956–75

Insurgency and Guerrilla Warfare

Under the right circumstances, guerrilla forces have achieved dramatic successes. In the 20th century, because of events in Russia, China and Vietnam, these successes have often been attributed to a particular ideology, but despite a relation to revolutionary theory, the campaigns point to the importance of pragmatic techniques and specific conditions. Guerrillas are dependent on certain types of terrain, the attitude of the civilian population, the will of the government being challenged, and the determination and skill of the individual guerrilla fighters. The essence of insurgent warfare is the 'hit-and-run' attack and concealment within the operational environment. T. E. Lawrence described the guerrilla army as resembling a gas – able to disperse as molecules to prevent a counter-strike, but also able to coalesce for its own operations. Yet, guerrilla warfare is, like terrorism (see **21**), actually evidence of comparative weakness: the insurgent lacks the manpower, equipment or support to defeat an enemy in conventional fighting and so turns to an asymmetrical campaign. Guerrilla forces seek to redress the imbalance of power by specific techniques: raids to cripple the capacity to wage war; inflicting casualties to demoralize an enemy; deliberately preventing a conclusion to a campaign in order to conduct a protracted war of attrition to exhaust resistance or rouse civilian opposition; securing external assistance for arms and supplies; focusing on the security of remote and inaccessible mountain or jungle fastnesses; or perhaps by disguising themselves within the fabric of civil society.[1]

24

Previous pages: Viet Cong insurgents moved through the civilian population like 'fish in the water' – here patrolling in a boat in 1961.

Guerrilla warfare requires a clear political agenda. Guerrillas must obtain the backing of the mass of the population for intelligence and recruits, but also for their own security from betrayal. It is often in their interests to create or sustain social-political unrest. They will attempt to offer an alternative form of government, since the ultimate aim of a guerrilla campaign is to secure political power over the territory in which they fight.

There have been a number of successful guerrilla campaigns throughout history, for example Alfred the Great's resistance to the Danes in 9th-century England, or Spanish opposition to French occupation in the Peninsular War (1808–14). However, although strategists since antiquity have thought about the techniques of guerrilla warfare, a definable theory of modern revolutionary warfare was only developed in the 20th century under the influence of the 1917 Russian Revolution. Vladimir Lenin's Bolsheviks perceived themselves as the élite revolutionary vanguard of highly politicized guerrillas who would lead the urban proletariat to establish a Marxist state.[2] Although reality in Russia did not come to match this theory, Communist success there inspired Marxists and revolutionaries around the world. Lenin's contribution to a theory of warfare was to insist on practical training and immersive political education.[3] He demanded utter loyalty, but acknowledged that only a very few would be suitable to form the centre cadre of the party-guerrilla force.

China, 1934–49

Mao Zedong and the Chinese Communists had almost been annihilated while following the Leninist model in 1927, and so developed their own version, based on the principles of the offensive and protracted war. This strategy was entirely located in rural areas among the mass of the population. In the 1950s, after the communist victory, groups outside China who sought to shake off colonial or élitist governments began to notice Mao's writings, the clarity and simplicity of which were attractive.

The vastness of China and the fact that 90 per cent of Chinese were peasants gave Mao two opportunities. Since he lacked military equipment, sound logistics and highly trained cadres of men (the Western standards of military effectiveness), he devised a theory that made maximum use of space, time and masses of people. Mao believed that since war was an extension of politics, the political agenda should dominate through an appeal to the people, and that military strength was directly proportional to the people's political commitment. Better-equipped armies sought to resolve guerrilla conflicts quickly; Mao therefore tried to prolong war so that the enemy would tire of the campaign politically and economically, and seek a resolution. Mao also managed to convince others of the 'inevitability' of communist victory.

Mao envisaged three, merging stages. The first was the mobilization and organization of the people in base areas, with indoctrination rather than guerrilla operations, and winning support by example, not by terror, as the priorities.[4] Mao's famous dictum was that guerrillas should be able to move among civilians like 'fish in the water'. In this phase, Mao was certainly helped

Mao Zedong addressing a crowd during the Long March. Political indoctrination was the first step in the mobilization of a revolutionary insurgent army in the Chinese Civil War.

by the preoccupation of the governing Nationalist authorities with the Japanese invasion in 1937. In stage two, Mao's fighters would strike their enemies in classic guerrilla operations. They would conduct ambushes and hit-and-run raids, but if attacked would refuse battle and fall back into the interior. The 'Long March' of 1934 was one such retreat, although it was much mythologized as a heroic, even offensive move. The idea was that additional areas should be overrun and developed, as in the first phase, as this would provide further recruits and permit the expansion of military units from the platoon and company size, to battalions of two or three companies. 'Regulars' would also form political regiments subjected to tougher discipline. By 1945, Mao had 14 base areas in China – the largest, on the Shansi-Hopei border, had a population of between 12 and 24 million. In the final stage, the guerrillas were strong enough to turn to conventional warfare.

Mao, like Clausewitz, understood that to achieve success guerrilla armies must either eventually reform into conventional forces, or work closely with regular troops. He also knew that the rural areas must engulf the towns and cities, and defeat the enemy in the field. The Vietnamese guerrilla leaders Ho Chi Minh and Vo Nguyen Giap adapted Mao's original ideas, adding 'confidence over their enemies' and 'terrorism' to the preparations for the final phase. In China, and later in Vietnam, guerrillas achieved their objectives, in part because of the mistakes made by their enemies.

Vietnam, 1956–75

Established in the 19th century by France, Indo-China was invaded by the Japanese in 1940. Aspirations for independence awakened by the arrival of this Asian power were soon dispelled, however, and there was bitter resentment when the Vichy French regime collaborated with the Japanese. Following liberation by the British in 1945, the wartime alliance and France's historic claims meant that the French soon re-occupied Indo-China. Almost immediately, nationalist, anti-colonial militias opened a guerrilla war, which expanded from small-scale skirmishes in 1948 to battalion-sized attacks on frontier forts in 1950. The Viet Minh guerrillas began to expand their network of bases, whereas French forces, unable to appeal to the local population to support them, were overextended by attempts to hold ground. Finally, the French attempted to lure the Viet commander Giap into a conventional engagement at Dien Bien Phu. As the culmination of this erroneous strategy, a force of French paratroops was surrounded and compelled to capitulate (March–May 1954).

The French defeat destroyed colonial authority, and the Americans stepped in to support the establishment of a 'free' South Vietnam against the new communist North Vietnam under Ho Chi Minh. Despite the promise of full democracy in the south, elections were not held in 1956 and the Americans became heir to the previous resented colonial powers. America, however, saw the issue in terms of the Cold War: communism had to be contained geographically to prevent a 'domino' collapse of Asia, and so it was increasingly sucked into a war to prevent the South Vietnamese army from being defeated by the North Vietnamese Army (NVA). NVA units operated in South Vietnam as guerrilla forces, using the Ho Chi Minh Trail in Cambodia and Laos as a covert supply line. The National Liberation Front, better known as the Viet Cong (VC), operated deeper in the south.[5] Although largely superseded by the NVA by 1965, the VC enjoyed successes with their simple hit-and-run tactics, dispersing into the civilian population after each attack.

The guiding principles of the NVA and Viet Cong followed the strategy of General Giap, broadly adopting Mao's three phases. The guerrillas first gained the support of the population through attacks on the government apparatus and 'armed propaganda'. Specially trained cadres of communist exiles from the south were sent back to ignite a 'people's war'. They stepped up attacks on the state's military forces and key institutions throughout the 1960s, and, in 1973–75, turned to conventional fighting to seize cities and overthrow the government. However, fighting did not follow a smooth linear pattern. Since it was impossible to defeat the Americans in open combat, the guerrillas' objective was to wear down the will of the Americans and South Vietnamese to continue resistance. But they suffered a number of setbacks and were forced to trade space for time – control of a particular area might have to be abandoned, and guerrilla forces dispersed, in order to stay operational.

Regardless of conditions and hazards, the North Vietnamese insurgents showed courage and determination in their campaign.

As Mao had stated: 'When the enemy advances, we retreat ... when the enemy retreats, we attack'.

The North Vietnamese Tet Offensive in 1968 broke the confidence of the public in both America and South Vietnam.[6] More than 100 urban centres were attacked by 80,000 insurgents, the American embassy in Saigon was raided, and there were pitched battles in the streets. From a tactical point of view the Tet Offensive was a costly failure, but it was a strategic success for the NVA – it failed to rouse the people of South Vietnam to rebellion,[7] but, despite defeat in bitter street battles, the attackers persuaded the United States that the conflict was 'unwinnable'. Previously the Americans had believed that it was only a matter of time before the North Vietnamese capitulated. After the Tet Offensive, although the American air campaign was stepped up, ground operations were scaled back and the number of US troops was gradually reduced in a process of 'Vietnamization'. By 1973, all American forces had been withdrawn, and the North Vietnamese offensive against the South began in earnest. It was concluded in 1975, with a return to conventional operations and a communist victory.

Analysis

Clearly the level of commitment of the guerrillas themselves, and their military training and expertise, are vital. As with regular forces, quality is better than quantity. Guerrillas, like conventional forces, are dependent on sound logistics and supplies. The image of the Viet Minh subsisting on a bowl of rice and a sip of water is a myth: fighters needed food and water, weapons and ammunition, mines and explosives, radio equipment and command decisions. They could tolerate austerity, and capture enemy supplies, but they could not function without support of this kind.

Control of supplies and resources can be the cause of guerrilla conflict, factional fighting and civil war. Today, the command of the drugs trade or oil resources are, according to historian Steven Metz, 'economic insurgencies'.[8] However, civilian support remains an essential component of the guerrillas' needs. Waging an information war is not enough: they must demonstrate that they are offering a better alternative, with a construction programme, reforms in land distribution or taxation, or by setting an example in sound and fair administration. Hearts and minds can be more important for both sides than firepower or occupation of the ground. Guerrilla warfare and counter-insurgency neatly illustrate that 'war', as Clausewitz reminded us, 'is an extension of politics'. Regardless of the tactical situation, political success almost certainly means operational success in the long term.

25

Malaya, 1948–60

Counter-Insurgency

Historically, the approach to dealing with rebellion was for official forces to crush the insurrection as rapidly as possible. Where an enemy refused to come to battle and deployed tactics of guerrilla warfare (see **24**), regular armies would often resort to the destruction of crops, livestock and property. With the fabric of their socio-economic power smashed, the rebels were compelled to gather and confront the state. The combination of military power in the hands of a ruler and this policy of state terror was often very effective. Asoka, the Mauryan emperor of India in the 3rd century BC, chose to crush the guerrilla resistance of Kalinga (on the borders of modern Bengal and Orissa) in a prolonged and brutal campaign.[1] On the site of the battle at the Daya River (mid-3rd century BC), stone stelas record those 100,000 Kalinga civilians and 10,000 of Asoka's warriors who perished. However, this campaign produced an unexpected outcome. The repression was so comprehensive that Asoka affirmed his commitment to Buddhism and for the remaining 40 years of his reign pursued a peaceful, conciliatory policy of governance. So it is here that we might identify the origins of the concept of 'hearts and minds', one of the most effective strategies of counter-insurgency.

25

Previous pages: Soldiers, bren guns at the ready, patrol a river and surrounding swamps in Malaya in 1957: painstaking work always characterizes the successful counter-insurgency.

Counter-insurgency – security measures set out as a coherent and integrated strategy against guerrillas – and counter-terrorism – a similarly interconnected range of security procedures against terror campaigns – have often relied on the overwhelming power of the state to crush the enemy; both are illustrated by Israeli dealings with the Palestinians (see p. 196). Specialist, loyal and often ruthless internal security forces have been a common element of state power, for example the paramilitaries of the French and Bolshevik revolutions. However, those dynasties and empires that won the loyalty of their subject populations, as well as overawing them, tended to last longer, with the Roman Empire as the greatest survivor.

The British Empire, too, spread its own ideology, including unfettered commercial enterprise, but preserved many existing local power structures and economic hierarchies, so that, while imperial troops were always quick to deal with disorder, the *modus operandi* was to avoid provoking resistance where possible. In the 20th century, with the evolution of democratic principles in British political life, the idea of government by consensus began to affect the colonies. When it became clear, after the Second World War, that the British Empire would have to transform itself from formal rule to more informal economic or political associations, state terror against anti-colonial forces was self-evidently excluded in terms of national and international expectations. The British counter-insurgency operations in the Malaya emergency (1948–60) were a crucial testing ground in this respect.

Malaya, 1948–60

In the Second World War, the British had trained and equipped the Malays to fight the Japanese. The Malay communists, banned before the war but dominant in the anti-Japanese struggle, were granted legal recognition after 1945 as a reward.[2] To further their aim of taking power, and using opposition to the idea of a Malaysian federation as a trigger for action, the communists began an insurgency in June 1948. They murdered British rubber planters and launched a campaign of strikes and disruption. The attacks by CTs (Communist Terrorists) escalated – against installations, transport systems, plantations and state personnel. Most of their recruits came from among the

500,000 ethnic Chinese Malays (Min Yuen), who tended to be poor and had little access to land, and did not have equal voting rights. Yet this racial factor was in fact a serious handicap. Despite the presence of a handful of ethnic Malays and Indonesians in the CT ranks, Malays regarded the British as allies, and likewise the British saw the Malays as 'friendlies'.

The CTs were organized into 'political regiments' with fighters, commissars, political sections, instructors, intelligence agents and propagandists. They tried to organize lectures and pamphlet campaigns among Malays, but with little success. However, they significantly boosted their prestige by assassinating the British High Commissioner, Sir Henry Gurney in 1951. They were keen to force the British to overreact, and to persuade the local population that the British were incapable of protecting them. As a result, their violence was often excessive: they rarely took prisoners, and tortured or killed survivors of their ambushes; pregnant women and children were mutilated. Isolated police posts were attacked, cafes bombed and trains machine gunned. By 1950, there were approximately 7,000 CTs raiding with impunity out of the Malayan jungles.

Responses to the Insurgency

The counter-insurgency was directed by Sir Gerald Templer, who combined the positions of civil governor and military commander. Realizing that a heavy-handed British response, epitomized by the shootings at Amritsar in India in 1919 (when British Indian troops had opened fire on a gathering of civilians, killing several hundred protestors), would be politically counter-productive, Templer resisted the temptation to send more troops into Malaya, or to over-react with a 'crackdown'.[3] There were never more than 35,000 men in theatre – although he was also constrained by the costs of a larger-scale counter-insurgency conflict, and by the need to avoid heavy casualties.

Templer's strategy was to isolate the CTs by a raft of measures. Rural populations were moved from their villages into collected and fortified settlements in the Briggs Plan inherited from his predecessor, Sir Harold Briggs. This forced move was resented at first, but Templer ensured that the villagers received money and that the new settlements has sufficient land to be viable, so that the people enjoyed economic benefits from the change. In this, the British had learnt from the South African War when many displaced civilians had died in camps through lack of such arrangements. In Malaya, medical care proved very popular.

Trackers from Borneo were brought in to Malaya to patrol in territory where insurgents were believed to be active. Identifying distinct and critical vulnerabilities can prove instrumental in defeating insurgents.

Special Forces (the SAS and the Malay Scouts – a local equivalent), consisting of experienced regular soldiers from the war years, together with contingents of the Brigade of Gurkhas, were deployed on patrols and ambushes. Sir Robert Thompson, Permanent Secretary for Defence in Malaya, had been a Chindit (a unit of Special Forces) in the Second World War, and he shared his experience of jungle warfare with senior officers in theatre. Patrols were gruelling and not all ambushes were sprung, but periodic successes became more frequent. Iban trackers from Borneo were also brought in to assist in locating the CTs.

The British were painstaking in their approach and reliant on local information. All ranks treated the locals as allies and cultivated the 'hearts-and-minds' ethos. Financial rewards were offered for information and the intelligence network built up a comprehensive picture of the personnel in CT units. The CTs were driven into the forests, away from population centres, and so deprived of supplies. These then had to be extorted from the Sakai tribesmen, who were thus alienated, a fact the security forces exploited – venturing out from their bases rendered the CTs vulnerable to detection and ambush.

Templer also understood the political aspect of the conflict: he immediately gave the Chinese the right to vote, put more Malays into positions of power and, most significantly, offered full independence once the insurgency was over. This both suited Britain's long-term plan to maintain close relations with former colonies for mutual economic benefit, and galvanized support from Malays. The last insurgents fled across the Thai border after independence, or surrendered in 1958.

Several local factors were significant. The CTs lacked the close proximity of an external ally for weapons and supplies (in contrast to the situation in Vietnam), while China was committed in Korea for part of the campaign. Operations were too remote for most journalists and there was little media coverage, in contrast to the accessibility and media 'popularity' of Vietnam. This meant that the communists lacked a platform to broadcast their message, and also that there was no criticism of British methods. Moreover, support for the communists was generally limited to ethnic Chinese. Malays had already fought against the Japanese alongside the British, a colonial power that had scarcely affected the Malays' way of life. There was therefore a degree of trust between the Malays and the British and Commonwealth allies. Furthermore, since Britain had demonstrated its commitment to peaceful transfers of power in India and Ceylon, the Malays had no reason to doubt the sincerity of Britain's decision to leave and welcomed their military power at the critical moment in their move towards independence.

So, the British had established at least part of a successful formula of counter-insurgency, adapted to the local circumstances. Although this approach was not applicable to every theatre – there were notable failures in Aden and Kenya – a set of measures had been established which could either neutralize guerrilla campaigns or 'hold the ring' until a political solution was reached. The political nature of the insurgent campaign was the first target, and the British 'outflanked' enemy goals with concessions in this field. Templer also knew that prosperity would deprive insurgents of support, hence the assistance for the new settlements. These were combined with an 'economy of effort' – a minimum military presence and limited application of force to achieve the object, sometimes known as a 'light footprint' – this reduced the potential for military-civilian tension as well as the costs, and hence avoided the operation becoming politically unacceptable. He established clear objectives, including coherent responsibilities for local authorities.

The central element of the campaign was winning the 'hearts and minds' of local civilians, especially through showing respect for their beliefs and values, which created conditions for maximum intelligence, good local recruitment and expanding political support.[4] The counter-insurgency operations themselves required the use of professional, well-trained troops rather than conscripts, but with permitted civilian authorities, including the police, leading wherever possible. A close working relationship between civil and military authorities was vital to restoring normal conditions of law and order.

The British also took the offensive. In several campaigns they dealt with the external suppliers of logistical support, intercepted communications and made the insurgents visible and vulnerable; they also exploited 'information war' methods to portray the enemy as criminal, encourage internal divisions and induce defections. They used deception wherever possible, such as irregular patrol patterns, to disrupt the CTs' plans.[5]

Analysis

In counter-insurgency operations Western powers have encouraged the maintenance of viable states; they exploit terrorist or insurgent vulnerabilities; seek ways to prevent the ideas spreading;[6] and identify means to dissolve such organizations.[7] There is a joint military-political agenda for these campaigns, conducted within the legal framework that most states have formulated to deal with terrorism.

Such campaigns, as in Malaya, can involve hard work over many years; indeed, the 21st-century 'war on terror' is known in military establishments as the 'long war'. In Malaya, there was a need for almost constant patrolling to reassure local populations, but also to locate, harass and defeat the insurgents. This was a painstaking process requiring great physical and mental endurance, and an expert grasp of survival or jungle warfare techniques to ensure that the rare encounters with guerillas delivered results.

As one veteran recalled:

> Mostly we were hacking our way through thick jungle with parangs or machetes, or flogging up steep, muddy jungle trails weighed down with packs and weapons and ammunition, or up to our hocks in swamps, from which we emerged well covered with leeches. It rained nearly every day in head-aching deluges and we were never dry ... and all this for about twenty-eight shillings a week [about £1.40 or $3].[8]

With all the tough conditions the troops had to endure, it was perhaps little comfort to them that counter-insurgencies rarely have a purely military solution. As one UN special representative put it: 'Soldiers in this situation do not win the peace – they simply hold the ring while a political solution is found.'[9]

Conclusion: How to Win on the Battlefield

Each chapter of this book has dealt with a tactical concept, with its execution illustrated by a range of historical examples. The final section of each chapter, the analysis, has evaluated how this concept – sometimes in conjunction with others – produced a particular outcome. So, although the themes have been selected and arranged individually, it is worth reiterating that a *combination* of tactics usually produced victory. Yet it would be wrong to assume that, in action, commanders simply selected ideas from a list for application. In many cases, tactical innovations may have been pure opportunism, and grander conceptions were often devised after the event to make sense of previous actions. Nevertheless, the papers and memoirs of some commanders do suggest that they had studied historical precedents and considered ways, in general terms, they might be used. Napoleon, for example, exhorted his subordinates to:

> Peruse again and again the campaigns of Alexander, Hannibal, Caesar, Gustavus Adolphus, Turenne, Eugene and Frederick. Model yourself upon them. This is the only means of becoming a great captain, and of acquiring the secret of the art of war. Your own genius will be enlightened and improved by this study, and you will learn to reject all maxims foreign to the principles of these great commanders.[1]

Previous pages: Russian troops in the streets of Stalingrad, 1942: they remind us that battles are won by discipline, endurance and motivation as much as by manoeuvre.

Mao Zedong wrote that such historical examples must be tested against personal experience: reading is important in learning, but so is application – perhaps even more so.[2] Decision-makers were crucial in all the examples we have mentioned. Sun Tzu and Clausewitz both recognized the centrality of leadership in producing victory. Sun Tzu urged patience and resolution as the best attributes for a commander and noted the frailties of human personality that most threatened disaster in war.[3] Clausewitz identified the 'genius' and intuition of leaders as being the means to overcome the inevitable deficiencies in battlefield intelligence that made rational calculations about manoeuvre impossible. Generals must possess strength of character, while avoiding obstinacy.[4] Field Marshal Montgomery argued that setting the objective and having the courage or will to see it through were the first priorities, and these had to be clearly communicated to subordinates. The commander must 'dominate events which encompass him' or risk losing the confidence of his men.[5] He had to create an 'atmosphere' based on inspiration, confidence, knowledge of the profession of arms and clear guidance. General George S. Patton put it more succinctly: 'Attack rapidly, ruthlessly, viciously, without rest, however tired and hungry you may be, the enemy will be more tired, more hungry. Keep punching.'[6]

Winning battles is, of course, crucial to generalship: men follow successful commanders with great loyalty. Exceptional military leaders possess a number of other qualities: the inspirational courage of Napoleon, the initiative and daring of Nelson, or the strength of will of Alexander. All need integrity and to foster a team spirit if men are to follow them freely.[7] In the 4th century BC, Xenophon proposed that commanders must have a 'wakeful eye' to logistics and the welfare of the troops, but also 'the knowledge of how to act ... the force of will and cunning to make them get the better of the enemy', and they should 'be trusted not to lead recklessly against the foe'.[8]

If wars and battles are fought by people, they are won or lost, to a large degree, in the 'hearts of men'.[9] The quality of the troops may be influenced by training, discipline and experience, and to some extent by their weapons, all of which contribute to the courage needed to endure battle, but ultimately what matters is morale and motivation. Montgomery believed the best troops were those who combined a strong fighting spirit with confidence in their leaders. In many cases a quiet, enduring determination was more valuable than the occasional episode of reckless courage. Field Marshal William Slim, who took command of the demoralized and defeated British Fourteenth Army in Southeast Asia during the Second World War, concentrated on restoring confidence, imposing strict discipline, insisting on efficiency, carrying out realistic training, and making sure that every man in the formation understood the importance of his role and how it fitted into the overall work of the army.[10]

Slim recognized that devotion to a cause started with self-respect and the bond of comradeship. When the army began to throw the Japanese back through Burma, the sense of achievement reinforced the foundations he had established. The rapidity of the advance towards Rangoon in 1945 meant that rations were reduced, and during a fierce Japanese counter-attack, Slim found himself behind a gun battery. The crews were soaked with sweat, heaving shells into the breach, striving to maintain a high rate of fire. When Slim apologized for the reduced rations during a pause in the fighting, a gunner replied: 'Never you mind about that, sir! Put us on quarter rations but give us the ammunition and we'll get you into Rangoon!'[11] The transformation of Slim's army had been complete.

Few generals with experience ever felt that speeches prior to going into action had much value, in spite of the classical texts' emphasis on it. Napoleon wrote:

> It is not set speeches at the moment of battle that render soldiers brave. The veteran scarcely listens to them, and the recruit forgets them at the first discharge. If discourses and harangues are useful, it is during the campaign; to do away with unfavourable impressions, to correct false reports, to keep alive a proper spirit in the camp, and to furnish materials and amusement for the bivouac.[12]

Commanders could, however, on rare occasions rally men who were wavering. Caesar grabbed a centurion's shield in a close battle against the Nervii; Frederick the Great inspired his men with a veiled criticism: 'Dogs!' he shouted, 'Would you live for ever?'[13] Lord Moran, who observed men's reactions to combat in the Second World War, felt courage was the will not to quit reaffirmed many times over.[14] He likened it to a bank account which could be drawn on, but which must be replenished and restored at every opportunity. He also believed it was a manifestation of good character. Slim identified an inspirational courage that occurred at the moment of crisis, but felt that an enduring courage was more common, and more valuable.[15] Napoleon stated: 'The first qualification of a soldier is fortitude under fatigue and privation. Courage is only the second: hardship, poverty and want are the best school for the soldier.'[16]

With highly motivated and well-trained troops, equipped with the appropriate weaponry, generals also need to be able to direct. Secure communications are a vital ingredient of success. To get troops to the right place, at the right time and with the right equipment, commanders also need to ensure that adequate resources, transport and logistics are available.

Success on the battlefield requires the techniques illustrated in this book, which can be grouped under the headings of the offensive, manoeuvre, active defence, the capacity to resist, psychology, and, if under severe pressure, the specific tactics of asymmetry. In conclusion, the factors that have ensured success on the battlefield through the ages may be summarized as follows:

- Adopt no single rigid approach or fixed doctrine; be flexible.
- Practise 'operational learning' to adapt and innovate quickly.
- Lead effectively – inspire, motivate, exploit the talents of subordinates.
- Identify the centre of gravity (strategy, key location or troops) and the moment when it changes; gather and disseminate intelligence and maintain tight security.
- Keep in mind strategic objectives to preserve the rational calculus of war: the means and the ends.

- Practise economy of effort – based on an assessment of risk and a cost-benefit analysis, know when to fight or withdraw, use an indirect approach, or trade space for time.
- Seize the initiative by taking the offensive, apply the principles of concentration (massing for local superiority), weight (of fire), shock and surprise, speed, and depth; maintain the tempo of operations and the momentum of attack, and keep up pressure on a retreating enemy.
- Manoeuvre one's own forces and the enemy into position, drawing the adversary into a killing area to annihilate his forces where possible; or outflank, envelop, enfilade or hit from behind. Adopt a strategic offence but a tactical defence.
- Create advantages by attrition, deception (false front, feint, camouflage, patterns, planted information and silhouettes), or a sudden change of tempo, to make the enemy strike into thin air.
- Break the enemy's will: create the idea of defeat in the mind of his commanders.
- Negotiate from a position of strength; know when to end a conflict.
- Practise information management and dominate decision-making (make the enemy appear to be in the wrong and isolate them, gather allies, demoralize the enemy and his population, disrupt and degrade his command and control, and get inside his decision-making cycle).
- Finally, accept the friction of war, and ensure you maintain a reserve.

Winning on the Battlefield: The Aims and Principles of War

1 Antoine-Henry, Baron de Jomini, *The Art of War*, translated by G. H. Mendell and W. P. Craighill, commentary by Horace Cocroft (Rockville, MD, 2006).

2 Carl von Clausewitz, *On War*, translated by Michael Howard and Peter Paret (Oxford, 2007); see also Michael I. Handel, *Masters of War: Classical Strategic Thought*, 3rd ed. (London and Portland, OR, 2001).

3 Sebastian Cox and Peter Gray (eds), *Air Power History: Turning Points from Kitty Hawk to Kosovo* (London and Portland, OR, 2002).

4 Colin S. Gray is sceptical of the threat in *Another Bloody Century: Future War* (London, 2005), pp. 221–22, 265. Paul Wilkinson believes the threat should be taken more seriously, in *Homeland Security in the UK: Future Preparedness for Terrorist Attack Since 9/11* (London, 2007).

5 See Vo Nguyen Giap, *How We Won the War* (Philadelphia, PA, 1977).

6 John Fuller to William Sloan, 2 July 1965. Fuller Papers 4/6/42/1. Liddell Hart Archive, King's College London.

1 The Attack at the Centre of Gravity

1 Montgomery of Alamein, *A History of Warfare* (London, 1968), p. 22.

2 Carl von Clausewitz, *On War* (Oxford, 2007) pp. 595–96. Clausewitz noted that the centre of gravity was usually located in the form of the enemy's army, but elsewhere he argued that the centre of gravity of *war* was the decisive battle (pp. 248 and 258). Baron de Jomini, by contrast, stated that the strategic centre of gravity of a *campaign* was the capital city of the enemy. However, in tactical terms, the centre of gravity was a specific location on the battlefield. Baron de Jomini, *The Art of War* (London, 1837), pp. 89–90.

3 See Denis Mack Smith, *Mussolini's Roman Empire* (London and New York, 1976).

4 G. P. B. Roberts, 'The Battle of Alam Halfa', in B. H. Liddell Hart, *History of the Second World War*, III (London, 1970), p. 1152.

5 For a critical appreciation of Montgomery's planning see John Ellis, *Brute Force: Allied Strategy and Tactics in the Second World War* (London and New York, 1990), pp. 279–84.

6 Michael Carver, 'Montgomery', in John Keegan (ed.), *Churchill's Generals* (London and New York, 1991), pp. 155–56.

7 Nigel Hamilton, *Monty: The Making of a General, 1887–1942*, I (London and New York, 1981), pp. 789–801.

8 Montgomery of Alamein, *El Alamein to the River Sangro; Normandy to the Baltic* (London, 1973), p. 31.

9 C. Barnett, *The Desert Generals* (London, 1960), p. 293.

10 In North Africa, the Allies demonstrated that firepower has to be mobile and concentrated, and able to support the advance. The Allied armour was a great mobile platform in the desert and able to operate at its maximum range. The open topography made for good 'tank country'. Infantry had the scope to defeat armour from prepared positions, but in the attack, the Allied infantry fulfilled their traditional function of taking and holding ground. The armour was able to take positions, but was less well suited to holding them unless dug in, 'hull down', like static artillery.

11 See Nigel Hamilton, *Monty*, 3 vols (London, 1981–86); John Keegan, *The Battle for History* (London, 1995), p. 60.

12 Montgomery, *A History of Warfare*, p. 23.

2 Counter-Attack

1 Niccolò Machiavelli, *The Art of War*, N. Wood (ed.), translated by E. Farneworth, 2nd rev. ed. (Indianapolis, 1965), Book IV, p. 655.

2 Ssu-ma Kuang (Sima Guang), *Chronicle of the Three Kingdoms*, I (Cambridge, MA, 1952).

3 B. H. Liddell Hart, *History of the First World War* (London, 1972), pp. 435–48.

4 Alexander Turner, *Cambrai, 1917: The Birth of Armoured Warfare* (London, 2007).

5 D. B. Nash, *The Imperial German Army Handbook, 1914–1918* (London, 1980), p. 76.

6 Ernst Jünger, *Storm of Steel*, translated by Michael Hofmann (London, 2003), p. 213.

7 Jack Horsfall and Nigel Cave, *Cambrai: The Right Hook* (Barnsley, 1999), p. 65.

8 C. E. Callwell, *Small Wars: Their Principles and Practice* (London, 1896), p. 202.

9 Callwell, *Small Wars*, p. 206.
10 Alan Axelrod, *Patton: A Biography* (New York, 2006).
11 See also Clausewitz, *On War* (Oxford, 2007), pp. 17–18.

3 Surprise Attack and Ambush

1 Tacitus, *Germania*, 6, cited in J. F. C. Fuller, *The Decisive Battles of the Western World*, I, rev. ed., edited by John Terraine (London, 1970), p. 170.
2 Adrian Murdoch, *Rome's Greatest Defeat: Massacre in the Teutoberg Forest* (Stroud, 2006).
3 *Dio's Roman History*, LVI, translated by Earnest Carey (London, 1916), p. 20.
4 Cassius Dio, *Roman History*, 61.21.2.
5 Cairo Radio, cited in Martin Gilbert, *Atlas of the Arab-Israeli Conflict* (London, 2002); Rais Ahmed Khan, *Propaganda in International Relations: A Case Study of Radio Cairo Broadcasts, 1956–59* (Winnipeg, 1968).
6 Chaim Herzog, *The Arab-Israeli Wars*, rev. ed. (London and New York, 2005) pp. 152–53.
7 Michael B. Oren, *Six Days of War: June 1967 and the Making of the Modern Middle East* (Oxford, 2002; New York, 2003), pp. 170–76.
8 Oren, *Six Days of War*, p. 180.
9 Jeremy Bowen, *Six Days. How the 1967 War Shaped the Middle East* (London, 2003), p. 223.
10 Bowen, *Six Days*, p. 221.
11 Ahron Bregman, *Israel's Wars: A History since 1947* (London and New York, 2002), p. 85.
12 Oren, *Six Days of War*, p. 287.

4 Envelopment and Double-Envelopment

1 See J. F. Lazenby, *Hannibal's War: A Military History of the Second Punic War* (Warminster, 1978; Norman, OK, 1998).
2 David Morgan, *The Mongols* (Oxford, 1986).
3 A. I. Akram, *The Sword of Allah: Khalid bin al-Waleed, His Life and Campaigns* (Lahore, 1969).
4 Anthony Beevor, *Stalingrad* (London and New York, 1998), p. 95; Edwin P. Hoyt, *199 Days: The Battle for Stalingrad* (London, 1993; New York, 1999), p. 87.
5 Ian Kershaw, *Hitler, II, Nemesis* (London, 2000), p. 407.
6 Hoyt, *199 days*, pp.189–92.
7 Beevor, *Stalingrad*, p. 165.
8 Kershaw, *Hitler, II*, pp. 543–45.

5 Flanking

1 See A. B. Bosworth, *Alexander and the East. The Tragedy of Triumph* (Oxford and New York, 1996).
2 Michael Howard, *The Franco-Prussian War: The German Invasion of France, 1870–71* (London and New York, 1961); J. F. C. Fuller, *The Decisive Battles of the Western World*, II (Stevenage, 1994), p. 256.
3 John H. Gill, *An Atlas of the 1971 India-Pakistan War* (Washington, DC, n.d., *c.* 2002), p. 55; Rob Johnson, *A Region in Turmoil: South Asian Conflicts since 1947* (London, 2005), pp. 158–59.
4 Gary W. Gallagher, *The American Civil War: The War in the East, 1861–1863* (London and Chicago, 2001), p. 64.
5 Gary W. Gallagher, *Chancellorsville: The Battle and its Aftermath* (Chapel Hill, NC, 1996).

6 Dominating the Terrain and Using the Environment

1 A. G. Lie, *The Inscriptions of Sargon II, King of Assyria* (London, 1929).
2 W. M. Mackenzie, *The Battle of Bannockburn, A Study in Medieval Warfare* (Glasgow, 1913; repr. Stevenage, *c.* 1989).
3 J. W. Croker, *The Croker Papers*, III (London, 1885), ch. 28.
4 H. L. Nevill, *Campaigns on the North West Frontier* (London, 1912), pp. 50–62.
5 Stephen Turnbull, *The Hussite Wars, 1419–1436* (London, 2004).
6 Fuller, *The Decisive Battles of the Western World*, p. 558.
7 Ludwig Reiners, *Frederick the Great*, translated by Lawrence Wilson (London, 1960), p. 185.
8 Both quoted in Fuller, *The Decisive Battles of the Western World*, p. 573.
9 *Histoire de mon temps: Frédéric le Grand* (Paris, 1879), p. 201.
10 *Histoire de mon temps*, p. 201.
11 Fuller, *The Decisive Battles of the Western World*, p. 577.

7 Echelon Attack

1 Rob Johnson, *The Iran-Iraq War, 1980–88* (London and New York, 2010).

8 Committing the Reserve

1 David Chandler, *The Campaigns of Napoleon* (London, 1993).

2 Christopher Duffy, *Austerlitz, 1805* (London, 1999).
3 Robert Johnson, *The Changing Nature of Warfare, 1792–1918: How War Became Global* (Taunton, 2002), pp. 91–92.

9 Blitzkrieg

1 The idea of infiltration by self-contained platoons was probably introduced initially by a French officer, Captain André Laffargue, in 1916. The German army took the idea and increased the strength of the units involved. John Terraine, *White Heat: The New Warfare 1914–1918* (London, 1982), p. 280.
2 Michael Geyer, 'German Strategy in an Age of Machine Warfare' in Peter Paret (ed.), *Makers of Modern Strategy: From Machiavelli to the Nuclear Age* (Oxford and Princeton, NJ, 1986).
3 Martin van Creveld, *Supplying War: Logistics from Wallenstein to Patton* (Cambridge and New York, 1977), pp. 142–45, 151–54.
4 Alastair Horne, *To Lose a Battle: France, 1940* (London and Boston, 1969).
5 John H. Eicher and David J. Eicher, *Civil War High Commands* (Stanford, CA, 2001).
6 David Walder, *The Short Victorious War: The Russo-Japanese Conflict, 1904–05* (London and New York, 1973), p. 257.
7 David M. Keithly and Stephen P. Ferris, '*Auftragstaktik* or Directive Control in Joint and Combined Operations', *Parameters: US Army War College Quarterly*, XXIX, 3 (1999), pp. 118–33.

10 Concentration of Firepower

1 Clive Bartlett, *The English Longbowman, 1330–1515* (London, 1995).
2 Geoffrey Wawro, *The Austro-Prussian War: Austria's War with Prussia and Italy in 1866* (Cambridge, 1996), p. 21.
3 We only have Roman sources for the campaign: Plutarch, in his biography of Crassus, and Cassius Dio (40.20), respectively written over 150 and 250 years after the disaster. The Roman account is that Crassus was betrayed by Abgar of Edessa, who urged him to march east across the desert rather than follow the line of the Euphrates in order to lure him towards the Parthian army commanded by Surenas; this may be true, although Alexander on a similar itinerary had also marched due east from Zeugma.
4 Bruce Vandervort, *Wars of Imperial Conquest in Africa* (London and Bloomington, IN, 1998), pp. 175–76.
5 Quoted in Thomas Pakenham, *The Scramble for Africa* (London and New York, 1991), p. 539.
6 Vandervort, *Wars of Imperial Conquest in Africa*, p. 177.
7 Pakenham, *The Scramble for Africa*, p. 545.
8 Izmat Zulfo, *Karari: The Sudanese Account of the Battle of Omdurman*, translated by P. Clark (London, 1980), p. 97.
9 Winston Churchill, *The River War*, II (London, 1899), p. 162.
10 Vandervort, *Wars of Imperial Conquest in Africa*, pp. 175–76.

11 Shock Action

1 Anna Comnena, *The Alexiad*, E. R. A. Sewter (ed.) (Harmondsworth, 1969), XIII:8, p. 416.
2 Ibn al-Qalanisi, 'Damascus Chronicle of the Crusades', in F. Gabrieli, *Arab Historians of the Crusades* (New York, 1989), p. 58.
3 Helen J. Nicholson (ed.), *Chronicle of the Third Crusade* (Aldershot, 1997), p. 234.
4 Nicholson, *Chronicle of the Third Crusade*, p. 233.
5 Baha al-Din Ibn Shaddad, *The Rare and Excellent History of Saladin*, D. S.Richards (Aldershot, 2002), p. 170.
6 Shaddad, *The Rare and Excellent History*, p. 175.
7 Richard to the Abbot of Clairvaux, 1 October 1191, in Peter W. Edbury (ed.) *The Conquest of Jerusalem and the Third Crusade* (Aldershot, 1996), p. 180.
8 John Selby, *The Thin Red Line of Balaclava* (London, 1970), p. 140.
9 Cecil Woodham-Smith, *The Reason Why* (London and New York, 1953), p. 229.
10 Winfried Baumgart, *The Crimean War, 1853–1856* (London, 1999), p. 128.

12 Co-ordination of Fire and Movement

1 Gary Sheffield, *Forgotten Victory. The First World War: Myths and Realities* (London, 2002), pp. 248–51.
2 Malcolm Brown, *1918: Year of Victory* (London, 1998), pp. 190–221.
3 Sheffield, *Forgotten Victory*, p. 239.
4 James Marshall-Cornwall, *Foch as Military Commander*

(London, 1972), p. 219.
5 Martin Marix Evans, *1918: The Year of Victories* (London, 2002), p. 160.
6 Brown, *1918: Year of Victory*, p. 209.
7 Cited in Sheffield, *Forgotten Victory*, p. 236.
8 Both in Sheffield, *Forgotten Victory*, p. 261.
9 Brown, *1918: Year of Victory*, pp. 254–55.

13 Concentration and Culmination of Force

1 Carl von Clausewitz, *On War* (Oxford, 2007), pp. 204–05.
2 John Terraine, *White Heat: The New Warfare, 1914–1918*, (London, 1982), p. 196.
3 B. H. Liddell Hart, *History of the First World War* (London, 1970), p. 459.
4 Terraine, *White Heat*, pp. 197–201.
5 B. H. Liddell Hart, *History of the Second World War* (London, 1970), p. 350.
6 E. Bauer, *The History of World War II* (Leicester, 1985), p. 249.
7 Liddell Hart, *History of the Second World War*, p. 351.

14 Seizing and Retaining the Initiative

1 Robert Marshall, *Storm from the East: From Genghis Khan to Kubiliai Khan* (London and Berkeley, CA, 1993).
2 Martin van Creveld, *Supplying War: Logistics from Wallenstein to Patton* (Cambridge, 1977), pp. 52 and 57.
3 David Chandler, *Waterloo: The Hundred Days* (London, 1980), p. 85.
4 Carl von Clausewitz, *On War* (Oxford, 2007), p. 203.
5 Richard H. Dillon, *The North American Indian Wars* (Greenwich, CT, 1997), p. 217.
6 Bruno Bauer, *The Second World War* (London, 1966), p. 514.
7 Tim Saunders, *Fort Eben Emael* (Barnsley, 2005).
8 Carl Shilleto, *Normandy: Pegasus Bridge and Merville Battery* (Barnsley, 1999).
9 Neil Barber, *The Day the Devils Dropped In: 9th Parachute Battalion in Normandy* (Barnsley, 2002).
10 Clausewitz, *On War*, p. 159.

15 Off-Balancing and Pinning

1 Montgomery of Alamein, *A History of Warfare* (London, 1968), p. 22.
2 Cited in J. F. C. Fuller, *The Decisive Battles of the Western World*, II, rev. ed., edited by John Terraine (London, 1970), pp. 81–82.
3 'Secret Memorandum', *Admiralty Committee Report* (cd 7120 of 1913), pp. 64–65.
4 Sir Nicholas Harris Nicholas (ed.), *The Despatches and Letters of Vice-Admiral Lord Nelson*, vol. VII (London, 1856), p. 60.
5 Nicholas, *Despatches and Letters*, vol. VII, p. 241.
6 Nicholas, *Despatches and Letters*, vol. VII, p. 241.
7 Cited in Fuller, *Decisive Battles of the Western World*, p. 90.
8 The Combined Fleet was pinned in place and destroyed, ship by ship. Only nine escaped; 17 were disabled, 13 were boarded (and subsequently sunk), and one was in flames. The French and Spaniards lost 4,400 killed, 2,500 wounded and around 4,000 were made prisoners. The British did not lose a single ship.
9 Fuller, *Decisive Battles of the Western World*, p. 88.

16 Mass

1 Richard Overy, *Why the Allies Won: Explaining Victory in World War II* (London and New York, 1995).
2 David French, *The British Way in Warfare, 1688–2000* (London, 1990), pp. 197–201; David Reynolds, Warren F. Kimball, A.O. Chubarian (eds), *Allies at War: The Soviet, American, and British Experience, 1939–1945* (New York and Basingstoke, 1994).
3 Curt Johnson and Mark McLaughlin, *Battles of the American Civil War* (Maidenhead, 1977), p. 108; David Nevin, *Sherman's March: Atlanta to the Sea (The Civil War)* (Alexandria, VA, 1986).
4 Gregory Jaynes, *The Killing Ground: Wilderness to Cold Harbor (The Civil War)* (Alexandria, VA, 1986), p. 56.
5 Jaynes, *The Killing Ground*, p. 64.
6 Jaynes, *The Killing Ground*, p. 92.
7 Douglas Freeman, *R. E. Lee: A Biography* (New York and London), p. 389.
8 Jaynes, *The Killing Ground*, p. 167.
9 Johnson and McLaughlin, *Battles of the American Civil War*, p. 118.
10 Johnson and McLaughlin, *Battles of the American Civil War*, pp. 149, 151.
11 Johnson and McLaughlin, *Battles of the American Civil War*, p. 154.

12 Jerry Korn, *Pursuit to Appomattox : The Last Battles* (*The Civil War*) (Alexandria, VA, 1986), pp. 88, 90.
13 Korn, *Pursuit to Appomattox*, p. 155.

17 Defence in Depth

1 Alan Clark, *Barbarossa: The Russo-German Conflict, 1941–45* (London, 1965), pp. 288–89.
2 Clark, *Barbarossa*, p. 294.

18 Strategic Offence and Tactical Defence

1 Satish Chandra, *Medieval India: From Sultanat to the Mughals* (Delhi, 2000), p. 28; John F. Richards, *The Mughal Empire* (Cambridge and New York, 1993).
2 Chandra, *Medieval India*, p. 29.
3 Paul K. Davis, *100 Decisive Battles: From Ancient Times to the Present* (Santa Barbara and Oxford, 1999), pp. 181–83.
4 Ahron Bregman, *Israel's Wars: A History Since 1947* (London and New York, 2002), p. 119.
5 S. el-Shazly, *The Crossing of Suez: The October War, 1973* (London, 1980), p. 122.
6 Bregman, *Israel's Wars*, pp. 134–37.
7 Bregman, *Israel's Wars*, p. 139.

19 Drawing the Enemy

1 Bernard Bachrach, *Early Carolingian Warfare: Prelude to Empire* (Philadelphia, PA, 2000).
2 We have many versions of this debate, but they are all coloured either by ignorance or the anxiety of one of the factions among the nobility to blame another for the subsequent disaster.
3 C. P. Melville and M. C. Lyons, 'Saladin's Hattin Letter' in B. Z. Kedar (ed.), *The Horns of Hattin: Proceedings of the Second Conference of the Society for the Study of the Crusades and the Latin East* (Jerusalem, 1992), pp. 208–12.
4 Ibn al-Athir, 'The Collection of Histories', in F. Gabrieli, *Arab Historians of the Crusades* (New York, 1957), p. 13.

20 Deception and Feints

1 Carl von Clausewitz, *On War* (Oxford, 2007), p. 203.
2 Baron de Jomini, *The Art of War* (London, 1837), p. 221.
3 Jon Latimer, *Deception in War* (London and Woodstock, NY, 2001), pp. 32–35; Iain R. Smith (ed.), *The Siege of Mafeking* (Johannesburg, 2001), pp. 161–63.
4 Horoshi Moriya and William Scott Wilson, *The 36 Secret Strategies of the Martial Arts* (Tokyo, 2008).
5 H. M. Le Fleming, *Warships of World War I* (London, 1967).
6 Sun Tzu, *The Art of War* (Oxford, 1971), p. 69.
7 Michael I. Handel, *Masters of War: Classical Strategic Thought*, 2nd ed. (London and Portland, OR, 1996), p. 94.
8 Roger Fleetwood Hesketh, *Fortitude: The D-Day Deception Campaign* (London and Woodstock, NY, 2000).
9 See Thaddeus Holt, *The Deceivers: Allied Military Deception in the Second World War* (London and New York, 2004).
10 Sun Tzu, *The Art of War*, p. 66.

21 Terror and Psychological Warfare

1 Farhad Daftary, *The Assassin Legends: Myths of the Isma'ilis* (London and New York, 1995), pp. 88–127.
2 Lloyd Pettiford and David Harding, *Terrorism: The Undeclared War* (New York, 2004) and *The New World War* (New York, 2003), citing Roger Scruton (1983), p. 546.
3 Stanley Sandler, *The Korean War* (London, 1999), pp. 204–07.

22 Attrition and Annihilation

1 Arden Bucholz, *Hans Delbrück and the German Military Establishment* (Iowa City, 1985).
2 J. H. Boraston (ed.), *Sir Douglas Haig's Despatches* (London, 1919 and 1979), p. 320.
3 General Erich von Falkenhayn, *General Headquarters 1914–1916 and its Critical Decisions* (London, 1919), p. 217. Richard Holmes notes there is some dispute about the veracity of Falkenhayn's claim to have planned in this way. Richard Holmes, *The Western Front* (London, 1999), pp. 82–83.
4 Alistair Horne, *The Price of Glory: Verdun 1916* (London and New York, 1962), p. 169.
5 Louis Barthas, *Les Carnets de Guerre de Louis Barthas, tonnelier, 1914–1918* (Paris, 1978).
6 Cited in Horne, *Price of Glory*.
7 Although attributed to Nivelle, because of its appearance in the Order of the Day (23 June 1916), the slogan had appeared among the troops before this date. Ian Ousby, *The Road to Verdun* (London, 2002), p. 231.

8 Cited in Horne, *Price of Glory.*
9 Holmes, *The Western Front*, p. 110; Rose E. B. Coombs, *Before Endeavours Fade*, 12th rev. ed. (Old Harlow, 2006).

23 Intelligence and Reconnaissance

1 Polybius, cited in Montgomery of Alamein, *A History of Warfare* (London, 1968), p. 16.
2 Sun Tzu, *The Art of War* (Oxford, 1971), p. 84.
3 Machiavelli, *The Art of War*, cited in Robert Greene, *The 33 Strategies of War* (London and New York, 2006), p. 166.
4 Baron de Jomini, *The Art of War* (Philadelphia, 1862), pp. 269–70.
5 Jeremy Black, *War Since 1990* (London, 2008).
6 F. H. Hinsley and Alan Stripp (eds), *Codebreakers: The Inside Story of Bletchley Park* (Oxford and New York, 1993).
7 F. H. Hinsley, *British Intelligence in the Second World War*, III (London, 1984).
8 Sun Tzu, *The Art of War*, p. 120.
9 Sun Tzu, *The Art of War*, p. 100.

24 Insurgency and Guerrilla Warfare

1 I. W. F. Beckett, *Modern Insurgencies and Counter-Insurgencies: Guerrillas and their Opponents since 1750* (London and New York, 2001).
2 Robert Service, *Lenin: A Biography* (London and Cambridge, MA, 2000), pp. 138–42.
3 Lewis H. Gann, *Guerrillas in History* (Stanford, CA, 1971).
4 Raymond F. Wylie, *The Emergence of Maoism: Mao Tse-tung, Ch'en Po-ta, and the Search for Chinese Theory, 1935–1945* (Stanford, CA, 1980).
5 Truong Nhu Tang, *A Viet Cong Memoir* (New York, 1985), chapter 7.
6 David L. Anderson, *The Vietnam War* (New York, 2005), p. 68.
7 Tran Van Tra, 'Tet' in Jayne S. Werner and Luu Doan Huynh (eds), *The Vietnam War: Vietnamese and American Perspectives* (New York, 1993), pp. 49–50.
8 Cited in Beckett, *Modern Insurgencies*, p. 237.

25 Counter-Insurgency

1 John Keay, *India: A History* (London, 2000), pp. 91–92.
2 Michael Dewar, *Brush Fire Wars: Minor Campaigns of the British Army since 1945* (London and New York, 1984), p. 29.
3 The Amritsar massacre caused a profound change in attitudes to the British in India. Keay, *India*, p. 476.
4 Noel Barber, *The War of the Running Dogs: How Malaya Defeated the Communist Guerrillas, 1948–60* (London, 1971).
5 David Kilcullen, *Counter Global Insurgency*, Nov. 2004, www.smallwarsjournal/documents/kilcullen1.pdf; see also 'Countering the Terrorist Mentality', *Foreign Policy Agenda*, 12, 5. www.usinfo.state.gov/pub/ejournalusa.htm (2007)
6 *Security, Terrorism and the UK*, Chatham House NSC Briefing Paper 05/01 (July, 2005).
7 UK Counter Terrorism Strategy, 2006.
8 Robin Neillands, *The British Empire: A Fighting Retreat, 1947–97* (London, 1996), p. 158.
9 Paddy Ashdown, reflecting on his service in Malaya. Paddy Ashdown, *Swords and Ploughshares: Bringing Peace to the 21st Century* (London, 2007), p. 3.

Conclusion: How to Win on the Battlefield

1 *Napoleon's Maxims of War*, translated by Lt Gen. G. C. D'Aguilar C.B. (Philadelphia, 1902).
2 Mao Zedong, *Selected Writings*, cited in Montgomery of Alamein, *A History of Warfare* (London, 1968), p. 19.
3 Sun Tzu, *The Art of War* (Oxford, 1971), pp. 78–79, 114–15, 136.
4 Carl von Clausewitz, *On War* (Oxford, 2007), p. 117.
5 Montgomery, *A History of Warfare*, p. 16.
6 Carlo D'Este, *Patton: A Genius for War* (London and New York, 1995).
7 Royal Military Academy Sandhurst, *Serve to Lead*, (Sandhurst, n.d.), pp. 22, 32.
8 Cited in *Serve to Lead*, p. 33.
9 Montgomery, *A History of Warfare*, p. 17.
10 Field Marshal William Slim, *Defeat Into Victory* (London, 1956), pp. 186–93.
11 Field Marshal William Slim, *Courage and Other Broadcasts* (London, 1957), cited in *Serve to Lead*.
12 *Napoleon's Maxims.*
13 Cited in *Serve to Lead*, p. 89.
14 Lord Moran, *Anatomy of Courage* (London, 2007).
15 Cited in *Serve to Lead*, p. 88.
16 *Napoleon's Maxims.*

Further Reading

General

Austin, N. J. E. and Rankov, N. B., *'Exploratio'. Military and Political Intelligence in the Roman World from the Second Punic War to the Battle of Adrianople* (London and New York, 1995).

Bennett, Matthew (ed.), *The Medieval World at War* (London and New York, 2009).

Black, Jeremy (ed.), *Great Military Leaders and their Campaigns* (London and New York, 2008).

Bradbury, Jim, *The Routledge Companion to Medieval Warfare* (London and New York, 2004).

Campbell, J. B., *War and Society in Imperial Rome, 31 BC–AD 284* (London and New York, 2002).

Clausewitz, Carl von, *On War*, translated by Michael Howard and Peter Paret (Oxford, 2007).

D'Este, Carlo, *A Genius for War: A Life of General George S. Patton* (London and New York, 1995).

Engels, Donald W., *Alexander the Great and the Logistics of the Macedonian Army* (Berkeley and London, 1978).

Fuller, J. F. C., *The Decisive Battles of the Western World and their Influence upon History*, rev. ed., edited by John Terraine, 2 vols (London, 1970).

Gilliver, C. M., *The Roman Art of War* (Stroud, 1999).

Goldsworthy, Adrian, *The Roman Army at War, 100 BC–AD 200* (Oxford and New York, 1996).

Hackett, John W. (ed.), *Warfare in the Ancient World* (London and New York, 1989).

Hanson, Victor Davis, *The Western Way of War. Infantry Battle in Classical Greece* (London and New York, 1989).

Hanson, Victor Davis (ed.), *Hoplites. The Classical Greek Battle Experience* (London and New York, 1991).

Hooper, Nicholas and Bennett, Matthew, *The Cambridge Illustrated Atlas of Warfare: The Middle Ages, 768–1487* (Cambridge and New York, 1996).

Howard, Michael, *War in European History* (Oxford and New York, 2009).

Jomini, Antoine-Henry Baron de, *The Art of War*, with an introduction by Charles Messenger (London and Novato, CA, 1992).

Lendon, J. E., *Soldiers and Ghosts. A History of Battle in Classical Antiquity* (New Haven, 2005).

Liddell Hart, Basil, *Strategy: The Indirect Approach*, 4th ed. (London, 1967).

Lloyd, Alan B. (ed.), *Battle in Antiquity* (London, 1996).

Paret, Peter (ed.), *Makers of Modern Strategy: From Machiavelli to the Nuclear Age* (Oxford and Princeton, NJ, 1986).

Sabin, Philip, van Wees, Hans and Whitby, Michael (eds), *The Cambridge History of Greek and Roman Warfare*, 2 vols (Cambridge, 2007).

Smail, R. C., *Crusading Warfare 1097–1193*, 2nd ed. (Cambridge and New York, 1995).

Souza, Philip de (ed.), *The Ancient World at War* (London and New York, 2008).

Strachan, Hew, *Carl von Clausewitz's* On War: *A Biography* (London and New York, 2007).

Sun Tzu, *The Art of War*, translated by Samuel B. Griffith (London, 1971).

1 The Attack at the Centre of Gravity
El Alamein, 1942

Carver, Michael, *El Alamein* (London and New York, 1962).

Hamilton, Nigel, *Monty: The Making of a General, 1887–1942*, I (London and New York, 1981).

Kitchen, Martin, *Rommel's Desert War: Waging World War II in North Africa, 1941–1943* (Cambridge and New York, 2009).

Montgomery of Alamein, *El Alamein to the River Sangro, Normandy to the Baltic* (London, 1973).

2 Counter-Attack
Cambrai, 1917

Military Operations France and Belgium, 1917, Cambrai, vol. 3 (London, repr. Imperial War Museum, 1992).

Nash, D. B. (ed.), *Imperial German Army Handbook, 1914–1918* (London, 1980).

Simkins, Peter, *The First World War: The Western Front, 1917–1918* (Oxford, 2002).

Wiest, Andrew, *The Western Front, 1917–1918. From Vimy Ridge to Amiens and the Armistice* (London, 2008).

3 Surprise Attack and Ambush
Teutoburg Forest, AD 9

Murdoch, Adrian, *Rome's Greatest Defeat: Massacre in the Teutoberg Forest* (Stroud, 2006).

Schlüter, W., 'The Battle of the Teutoburg Forest: Archaeological Research at Kalkriese near Osnabrück', in *Roman Germany. Studies in Cultural Interaction*, Journal of Roman Archaeology, Supplementary Series no. 32 (Portsmouth/Rhode Island, 1999), pp. 125–59.

Wells, Peter S., *The Battle that Stopped Rome: Emperor Augustus, Arminius, and the Slaughter of the Legions in the Teutoburg Forest* (New York, 2003).

Whittaker, C. R., *Frontiers of the Roman Empire: A Social and Economic Study* (Baltimore, 1994).

The Six Day War, 1967

Bowen, Jeremy, *Six Days: How the 1967 War Shaped the Middle East* (London, 2003).

Bregman, Ahron, *Israel's Wars: A History since 1947* (London and New York, 2002).

Herzog, Chaim, *The Arab-Israeli Wars*, rev. ed. (London and New York, 2005).

Oren, Michael B., *Six Days of War: June 1967 and the Making of the Modern Middle East* (Oxford, 2002; New York, 2003).

4 Envelopment and Double-Envelopment

Walaja, 633

Frye, R. N. (ed.), *The Cambridge History of Iran*, vol. 4, (London, 1975).

Muir, Sir William, *The Caliphate: Its Rise, Decline and Fall from Original Sources* (London, 1891; repr. 2004).

Operation Uranus, 1943

Beevor, Anthony, *Stalingrad* (London and New York, 1998).

Glantz, David M., *Armageddon in Stalingrad: September–November 1942*. (Lawrence, KA, 2009).

Overy, Richard, *Russia's War* (London, 1999).

Roberts, Geoffrey, *Victory at Stalingrad* (Harlow, 2002).

5 Flanking

Bouvines, 1214

Duby, G., *The Legend of Bouvines: War, Religion and Culture in the Middle Ages*, translated by C. Tihanyi (Cambridge and Berkeley, CA, 1990).

France, John, *Western Warfare in the Age of the Crusades, 1000–1300* (London and Ithaca, NY, 1999).

Verbruggen, J. F., *The Art of Warfare in Western Europe during the Middle Ages*, translated by S. Willard and R. W. Southern (Woodbridge and Rochester, NY, 1997).

Chancellorsville, 1863

Johnson, Curt and McLaughlin, Mark, *Battles of the American Civil War* (Maidenhead, 1977); *Battles of the Civil War* (New York, 1977).

Chandler, David and Smith, Carl, *Chancellorsville, 1863* (London, 1998).

Earle, Peter, *Robert E. Lee* (London and New York, 1973).

Selby, John, *Stonewall Jackson as Military Commander* (London and Princeton, 1968).

Stackpole, Edward J., *Chancellorsville: Lee's Greatest Battle* (Harrisburg, 1958).

6 Dominating the Terrain and Using the Environment

Horice, 1423

Bradbury, Jim, *The Routledge Companion to Medieval Warfare* (London and New York, 2004).

Hooper, Nicholas and Bennett, Matthew, *The Cambridge Illustrated Atlas of Warfare: The Middle Ages, 768–1487* (Cambridge and New York, 1996).

Leuthen, 1757

Duffy, Christopher, *The Army of Frederick the Great* (London, 1974; Chicago, 1996).

Millar, Simon, *Rossbach and Leuthen, 1757: Prussia's Eagle Resurgent* (Oxford, 2002).

Showalter, Dennis, *The Wars of Frederick the Great* (London and New York, 1996).

7 Echelon Attack

Leuctra, 371 BC

Buckler, John, *The Theban Hegemony 371–362 BC* (Cambridge, MA, 1980).

Tuplin, C. J., 'The Leuctra Campaign: some outstanding problems', *Klio* 69 (1987), pp. 72–107.

Hanson, Victor Davis, 'Epaminondas, the Battle of Leuctra (371 BC), and the "Revolution" in Greek Battle Tactics', *Classical Antiquity* 7 (1988), pp. 190–207.

8 Committing the Reserve

Strasbourg, 357

Nicasie, M. J., *Twighlight of Empire. The Roman Army from the Reign of Diocletian until the Battle of Adrianople* (Amsterdam, 1998).

Elton, Hugh, *Warfare in Roman Europe, AD 350–425* (Oxford and New York, 1996).

Austerlitz, 1805

Castle, Ian, *Austerlitz, 1805: The Fate of Empires* (Oxford, 2002; Westport, CT, 2005).

Duffy, Christopher, *Austerlitz, 1805* (London, 1999).

Esdaile, Charles J., *The Wars of Napoleon* (Harlow and New York, 1995).

Hourtoulle, François-Guy, *Austerlitz: The Empire at its Zenith* (Paris, 2008).

Rothenberg, G. E., *The Napoleonic Wars* (New York, 2005).

9 Blitzkrieg

Khalkin Gol, 1939; Operation August Storm, Manchurian Campaign, 1945

Chaney, Otto P., *Zhukov* (Norman, OK, 1996).

Glantz, David M., *Soviet Operational and Tactical Combat in Manchuria, 1945, August Storm* (London and Portland, OR, 2003).

Reese, Roger, *The Soviet Military Experience: A History of the Soviet Army, 1917–1991* (London and New York, 2000).

Spahr, William J., *Zhukov: The Rise and Fall of a Great Captain* (Novato, CA, 1993).

10 Concentration of Firepower

Carrhae, 53 BC

Goldsworthy, Adrian, *The Roman Army at War, 100 BC–AD 200* (Oxford and New York, 1996), pp. 60–68.

Sherwin-White, A. N., *Roman Foreign Policy in the East 168 BC to AD 1* (London and Norman, OK, 1984).

Omdurman, 1898

Featherstone, Donald, *Omdurman 1898: Kitchener's Victory in the Sudan* (London, 1993; Westport, CT, 2005)

Vandervort, Bruce, *Wars of Imperial Conquest in Africa, 1830–1914* (London and Bloomington, IN, 1998).

11 Shock Action

Arsuf, 1191

Gillingham, John, *Richard I* (New Haven and London, 1999).

France, John, *Western Warfare in the Age of the Crusades, 1000–1300* (London and Ithaca, NY, 1999).

Smail, R. C., *Crusading Warfare, 1097–1193*, 2nd ed. (Cambridge and New York, 1995).

Verbruggen, J. F., *The Art of Warfare in Western Europe during the Middle Ages*, translated by S. Willard and R. W. Southern (Woodbridge and Rochester, NY, 1997).

Balaclava, 1854

Adkin, Mark, *The Charge* (London, 2004).

Dutton, Roy, *Forgotten Heroes: The Charge of the Heavy Brigade* (Wirral, 2008).

Kelly, Christine (ed.), *Mrs Duberly's War: Journals and Letters from the Crimea, 1854–56* (Oxford, 2008).

Sweetman, John, *The Crimean War* (Oxford, 2001).

12 Co-ordination of Fire and Movement

Cerignola, 1503

Oman, Charles, *A History of the Art of War in the Sixteenth Century* (London, 1987; original ed. 1937).

The Hindenburg Line, 1918

Griffith, Paddy, *Battle Tactics on the Western Front: The British Army's Art of Attack, 1916–18* (New Haven and London, 1994).

Neiberg, Michael S., *Foch: Supreme Allied Commander in the Great War* (Dulles, 2003).

Oldham, Peter, *The Hindenburg Line, 1918* (London, 1997).

Sheffield, Gary, *Forgotten Victory. The First World War: Myths and Realities* (London, 2002).

Travers, Tim, *How the War was Won: Factors That Led to Victory in World War One* (Barnsley, 2005).

13 Concentration and Culmination of Force

Jagdgeschwader, the Western Front, 1916–17

Neumann, Georg Paul, *The German Air Force in the Great War*, translated by J. E. Gurdon (London, repr. 2009).

vanWyngarden, Greg, *Richthofen's Flying Circus: Jagdgeschwader Nr 1* (Oxford, 2004).

Werner, Johannes, *Knight of Germany: Oswald Boelcke German Ace* (Newbury, 2009).

Midway, 1942

Isom, Dallas Woodbury, *Midway Inquest: Why the Japanese Lost the Battle of Midway* (Bloomington, IN, 2007).

Smith, Peter C., *Midway, Dauntless Victory: Fresh Perspectives on America's Seminal Naval Victory of World War II* (Barnsley, 2007).

14 Seizing and Retaining the Initiative

Eben Emael, 1940

Dunstan, Simon, *Fort Eben Emael: The Key to Hitler's Victory in the West* (Oxford, 2005).

Saunders, Tim, *Fort Eben Emael* (Barnsley, 2005).

Pegasus Bridge, 1944

Shilleto, Carl, *Normandy: Pegasus Bridge and Merville Battery* (Barnsley, 1999).

15 Off-Balancing and Pinning

Trafalgar, 1805

Hayward, Joel, *For God and Glory: Lord Nelson and His Way of War* (Annapolis, MD, 2003).

Hoock, Holger, *History, Commemoration and National Preoccupation: Trafalgar, 1805–2005* (Oxford, 2007).

Knight, Roger, *The Pursuit of Victory: The Life and Achievement of Horatio Nelson* (London and New York, 2005).

Lambert, Andrew, *Nelson: Britannia's God of War* (London, 2005).

16 Mass

The Overland Campaign, 1864–65

Fuller, J. F. C., *The Generalship of Ulysses S. Grant* (London and New York, 1929).

Reid, Brian Holden, *The Civil War and the Wars of the Nineteenth Century* (London, 2002).

King, Curtis S. et al., *Staff Ride Handbook for the Overland Campaign, Virginia, 4 May to 15 June 1864 : A study in Operational - Level Command* (Fort Leavenworth, 2006).

17 Defence in Depth

Alesia, 52 BC

Gilliver, Kate, *Caesar's Gallic Wars, 58–50 BC* (Oxford, 2002).

Holmes, T. Rice, *Caesar's Conquest of Gaul*, 2nd ed. (Oxford, 1911).

Kursk, 1943

Glantz, David M. and House, Jonathan M., *The Battle of Kursk* (Lawrence, KS, 1999).

Glantz, David M., *After Stalingrad: The Red Army's Winter Offensive, 1942–1943* (Solihull, 2009).

Healy, Mark, *Zitadelle: The German Offensive Against the Kursk Salient, 4–17 July 1943* (Staplehurst, 2008).

18 Strategic Offence and Tactical Defence

Panipat, 1526

Gommans, Jos, *Mughal Warfare: Indian Frontiers and Highroads to Empire, 1500–1700* (London and New York, 2002).

Richards, John F., *The Mughal Empire* (Cambridge and New York, 1996).

Yom Kippur, 1973

Herzog, Chaim, *The Arab-Israeli Wars*, rev. ed. (London and New York, 2005).

Herzog, Chaim, *The War of Atonement: The Inside Story of the Yom Kippur War* (London, 1975).

19 Drawing the Enemy

Hattin, 1187

Kedar, B. Z., 'The battle of Hattin revisited', in B. Z. Kedar (ed.), *The Horns of Hattin: Proceedings of the Second Conference of the Society for the Study of the Crusades and the Latin East* (Jerusalem, 1992).

Lyons, M. C. & Jackson, D. E. P., *Saladin. The Politics of Holy War* (Cambridge, 1982).

Riley-Smith, Jonathan (ed.), *The Atlas of the Crusades* (London, 1991).

Smail, R. C., *Crusading Warfare 1097–1193*, 2nd ed. (Cambridge, 1995).

Napoleon in Russia, 1812

Austin, Paul Britten, *1812: Napoleon's Invasion of Russia* (London, 2000).

Connelly, Owen, *Blundering to Glory: Napoleon's Military Campaigns*, 3rd ed. (Landham, MD, and Oxford, 2006).

Mikaberidze, A., *The Russian Officer Corps in the Revolutionary and Napoleonic Wars, 1792–1815* (El Dorado, 2005).

Parkinson, Roger, *Fox of the North: The Life of Kutuzov* (New York and London, 1976).

20 Deception and Feints

Kurikara, 1183

Shively, Donald H. and McCullough, William H., *The Cambridge History of Japan*, Vol. 2 (Cambridge and New York, 1999).

Turnbull, Stephen R., *The Samurai: A Military History* (London and New York, 1996).

Q-Ships, 1915–17

Bridgland, Tony, *Sea Killers in Disguise: The Story of the Q-Ships and Decoy Ships in the First World War* (London and Annapolis, MD, 1999).

Lake, Deborah, *Smoke and Mirrors: Q-Ships Against the U-Boats in the First World War* (Stroud, 2006).

21 Terror and Psychological Warfare

Thebes, 335 BC

Bosworth, A. B., *A Historical Commentary on Arrian's History of Alexander*, Vol. I (Oxford, 1980), pp. 73–84.

Fuller, J. F. C., *The Generalship of Alexander the Great* (London, 1958).

Palestinian Terrorism, 1950–99

Rogan, Eugene L. and Shlaim, Avi (eds), *The War for Palestine: Rewriting the History of 1948*, 2nd ed. (Cambridge and New York, 2007).

Gelvin, James L., *The Israel-Palestine Conflict: One Hundred Years of War*, 2nd ed. (Cambridge and New York, 2007).

Herzog, Chaim, *The Arab-Israeli Wars*, rev. ed. (London and New York, 2005).

Laqueur, Walter, *The Age of Terrorism* (Boston, 1987).

22 Attrition and Annihilation

Verdun, 1916

Brown, Malcolm, *Verdun, 1916* (Stroud, 1999).

Horne, Alistair, *The Price of Glory: Verdun, 1916* (London and New York, 1962; new ed. London, 2007).

Ousby, Ian, *The Road to Verdun* (London, 2002)

23 Intelligence and Reconnaissance

The Battle of the Atlantic, 1941–45; Cape Matapan, 1941; North Cape, 1943

Hinsley, F. H. and Stripp, Alan (eds), *Codebreakers: The Inside Story of Bletchley Park* (Oxford and New York, 1993).

Lavery, Brian, *Churchill's Navy: The Ships, Men and Organisation, 1939–1945* (London, 2006).

Winton, John, *Cunningham: The Greatest Admiral Since Nelson* (London, 1998).

24 Insurgency and Guerrilla Warfare

China, 1934–49

Mao Tse-Tung [Mao Zedong], *On Guerrilla Warfare*, translated by Samuel B. Griffith (Urbana, 1961; new ed. 2000).

Taber, Robert, *War of the Flea. The Classic Study of Guerrilla Warfare* (New York, 1965; repr. Washington, DC, 2002).

Vietnam, 1956–75

Giap, Vo Nguyen, *People's War, People's Army: The Viet Cong Insurrection Manual for Underdeveloped Countries* (Honolulu, 2001).

Kolko, Gabriel, *Vietnam: Anatomy of a War, 1940–1975* (London, 1985).

25 Counter-Insurgency

Malaya, 1948–60

Beckett, Ian F. W., *Modern Insurgencies and Counter-Insurgencies: Guerrillas and their Opponents since 1750* (London and New York, 2001).

Jackson, Robert, *The Malayan Emergency: The Commonwealth's Wars 1948–1966* (London and New York, 1991).

Nagl, John, *Learning to Eat Soup with a Knife: Counterinsurgency Lessons from Malaya and Vietnam* (London and Chicago, 2005).

The U.S. Army/Marine Corps Counterinsurgency Field Manual (Chicago, 2007).

Sources of Illustrations

All battle plans: Red Lion Prints

akg-images 11, 21, 26, 62–63, 78–79, 114, 148, 158–159, 176–77, 182, 222
akg-images/British Library 57, 169
Bibliothèque Municipale, Amiens 66
The Art Archive 200–01
Gernsheim Collection, Austin, Texas 156
Staatliche Museen zu Berlin 74
National Army Museum, Camberley 82
Bettman/Corbis 16, 102–03, 122–23, 197
Corbis 2–3, 30, 91, 140
Forestier Yves/Corbis/Sygma 139
Mahar Attar/Corbis/Sygma 192–93
Hulton-Deutsch Collection/Corbis 34–35, 38, 132–33
Michael Nicholson/Corbis 179
David Rubinger/Corbis 172
Roger Fenton 109
Getty Images 20, 41, 94–95, 228–29, 232
AFP/Getty Images 166–67
Popperfoto/Getty Images 18–19
Hirmer 81, 97
Inter Film Services: 112–13
Museum and Park, Kalkriese 36
Neue Galerie, Kassel 68
British Museum, London 106
Imperial War Museum, London 15, 50–51, 116–17, 120, 130, 137, 184–85, 189, 210–11, 212, 215, 218, 236
National Maritime Museum, London 145
Tate, London 142–43
Museo Nazionale Archeologico, Naples 194
Bibliothèque Nationale de France, Paris 10
Musée d'Orsay, Paris 12, 152
ECPAD Ministère de la Defence, Port d'Ivry 204
Private Collection 127, 155
Museum of the Confederacy, Richmond, Va. 59
Topham Picture Point/TopFoto.co.uk 220–21
Tran Binh Khuoi 226
U.S. Army Military History Institute, Carlisle Barracks, Pa. 60
Ullstein Bilderdienst 86–87
I Musei Vaticani 8
Roger-Viollet 206
Library of Congress, Washington, D.C. 186
National Archives, Washington, D.C. 54–55, 150–51
Georg Zelma 44

Index

Page numbers in *italic* indicate illustrations

Index